On General Relativity

On General Relativity

An Analysis of the Fundamentals of the Theory of General Relativity and Gravitation

By

A. Mercier (Berne – Switzerland)
H.-J. Treder (Berlin – GDR)
W. Yourgrau (Denver – USA)

With 3 Figures

Akademie-Verlag · Berlin 1979

Erschienen im Akademie-Verlag, DDR – 108 Berlin, Leipziger Straße 3–4
Lektor: Dipl.-Phys. Ursula Heilmann

Lizenznummer: 202 · 100/237/78
Gesamtherstellung: IV/2/14 VEB Druckerei »Gottfried Wilhelm Leibniz«,
445 Gräfenhainichen · 5148
Umschlaggestaltung: Karl Salzbrunn
Bestellnummer: 762 6043 (6505) · LSV 1115
Printed in GDR
DDR 22,– M

TABLE OF CONTENTS

1. ORIGIN AND PROBLEMATICS OF GRG

1.1. Introduction

The theory of General Relativity and Gravitation is always thought of in connection with Gravitation. We have created the abreviation GRG for *General Relativity and Gravitation* (in French, in the opposite order: *Gravitation et Relativité Générale*). One uses too, the word gravitodynamics, to be compared with electrodynamics but not with hydrodynamics: this is due to the fact that "gravito", "electro" ... refer to distinct kinds of interaction, "hydro" or "aero" e.g. to states of matter. In both cases, "dynamics" designates a theory about reversible phenomena connected either with the interaction under consideration or the realised state. When one talks about thermodynamics one does not mean a dynamics in the same sense but a (still rudimentary) theory of irreversibility which is superposed upon the existence of interactions as well as upon the possibility of various states of matter.

Now, at least at the origin, Newtonian mechanics was, too, and in principle, even uniquely a theory of gravitation, i.e. of a fundamental interaction. In that *first sence*, GRG will be a *generalization of Newtonian mechanics*.

However, in the mind of Einstein's his concern with that particular generalization was only mediate. For, he had originally elaborated a new kinematics of relative motions restricted to Galilean coordinate systems, a kinematics which strictly speaking applies to electrodynamics only but not to gravitodynamics as the latter was conceived within Newtonian mechanics. So, if the generalization was, on the one hand, intended to describe gravitation, it was, on the other hand, from this point of view, intended to get rid of the restriction imposed by considering Galilean systems only. In that *second sense, GRG is a generalization of Lorentzian kinematics*, i.e. of the theory of special relativity. GRG is unification of gravitodynamics and special relativity by generalization of both theories.

The Newtonian mechanics of gravitation rests on two premises of its own:
(a) a postulate and (b) a law
which can be enunciated in one of the following forms:

	equation of motion	*interaction law*
original Newtonian form	$\frac{\mathrm{d}\vec{p}}{\mathrm{d}t}=\vec{F}$	$\vec{F}=-G\frac{Mm}{r^3}\vec{r}$
semi modern Newtonian form by means of a (pseudo-)field	$\frac{\mathrm{d}\vec{p}}{\mathrm{d}t}=-\nabla\varphi$	Poisson's equation: $\Delta\varphi=-4\pi G\varrho$
by means of a Lagrangian function L	$\delta\int L\mathrm{d}t=0$	$L=\frac{mv^2}{2}-m\varphi$
by means of a Lagrangian density $\mathfrak{L}$	$\delta\int\mathfrak{L}\mathrm{d}\tau=0$	$\mathfrak{L}=\frac{\mu v^2}{2}-\mu\varphi$
canonical form	$\delta\int(pq-H)\,\mathrm{d}t=0$	$H=\frac{p^2}{2m}+m\varphi$

as well as other forms.

Special relativistic kinematics rests on one single premise, the Lorentz-transformation

$$(\mathbb{L}):(\vec{r},t)\rightarrow(\vec{r}',t')=\mathbb{L}(\vec{r},t).$$

Hence, one can expect to deal with a generalization of either ($\mathbb{L}$) or (a) or (b). This is indeed the way Einstein followed. However, the generalization he gave for the law under (b) has been such that the equation of motion was finally found to be redundant, i.e. in the new theory it lost its independence. Yet, he did not notice it from the beginning. The whole state of affairs related to the famous "principle" of equivalence is implied there just in the opposite way of what happens in Newtonian mechanics.

Newtonian mechanics is essentially a dynamics, Einsteinian "mechanics" is essentially a kinematics. In Newtonian dynamics, *forces of inertia* are called "forces" and are looked upon as such entities, as if they were felt as forces; but we know that they are not real since they only arise form relative motions. In (general) Einsteinian kinematics, exactly the opposite takes place: all the forces are forces of inertia, they are kinematical effects, there are no real forces and consequently it must always be possible to find a coordinate system in which (at least momentarily and locally) the motion is not accelerated.

However, there is no hope to find this coordinate system by mere Lorentzian transformations since they will never eliminate accelerations.

To reach this aim, one has therefore to conceive of *sufficiently* general coordinate transformations, of which ($\mathbb{L}$) will have to be a particular case or an approximation.

Now, ($\mathbb{L}$) transforms varieties of the Minkowski space $\mathbb{E}^4$ into one another, that is, pseudo-Euclidean varieties whose metric is given by

$$\mathrm{d}s^2 = c^2\mathrm{d}t - (\mathrm{d}x^2 + \mathrm{d}y^2 + \mathrm{d}z^2) = \sum_{i=0}^{3} (\mathrm{d}x^i)^2 = \sum_{i=0}^{3} (\mathrm{d}x'^i)^2 .$$

These varieties may be osculatory to varieties of a metric space called the Riemannian space $\mathbb{R}^4$, having the same number of dimensions and admitting of null distances too. ($\mathbb{L}$) will then have but a local meaning, i.e. valid in an infinitesimal or differential neighbourhood, which implies that the components of the matrix which will have to replace ($\mathbb{L}$) be in general functions of the coordinates. These coordinates y^i will be called curvilinear. Pythagoras' theorem will in general not read in its simple form but

$$\begin{aligned} \mathrm{d}s^2 &= g_{ik}\mathrm{d}y^i\mathrm{d}y^k, \qquad \text{where} \qquad g_{ik} = g_{ik}(y), \quad y = \{y^1, y^2, y^3, y^4\} \\ &= g'_{ik}\mathrm{d}y'^i\mathrm{d}y'^k, \qquad\qquad\qquad g'_{ik} = g'_{ik}(y') . \end{aligned}$$

Transformations $y \to y'$, $g_{ik} \to g'_{ik}$ will have to be assumed, implying the whole of tensor calculations expressing the *variance* of *tensors*.

The (covariant) tensor g_{ik}, called *metric tensor*, determines the variety.

For what physical reasons such and such Riemannian variety should be realized rather than another one? That is the most fundamental question which Einstein had to put himself.

Now, special relativity, for which in an orthogonal Cartesian coordinate system x^i

$$(g_{ik}) = \begin{pmatrix} 1 & 0 & 0 & 0 \\ 0 & -1 & 0 & 0 \\ 0 & 0 & -1 & 0 \\ 0 & 0 & 0 & -1 \end{pmatrix} = (\overset{\circ}{g}_{ik})$$

applies in the absence of gravitation only, or still better under the assumption of the sole existence of an electromagnetic field in which charged material points or extentionless magnetic dipoles circulate, in other words, in the absence of sources of the "potential" or "pseudo-field" φ. In the ratio $\frac{e}{m}$ of charge to mass, m can only assume the significance of an inertial mass and not of a heavy mass.

But masses are heavy, grave. Dropping provisionally the tricky discussion concerning the influence of the mass $\frac{W}{c^2}$ attached to any distribution of electromagnetic energy $W = \frac{\varepsilon}{2}(E^2 + B^2)\,dV$ [1], one may guess that the g_{ki} will differ from the g_{ik} *because of the distribution of mass.*

Of course, from the point of view of general coordinates y^i, the expresion of a mass distribution will imply not only a density of mass $\mu(y)$ but also a whole tensorial complex. The latter is the *material energy-momentum tensor* M^{ik}—where "material" is meant to distinguish it from energy-momentum contributions due for instance to the electromagnetic field.

The g_{ik} are of course functions of the M^{ik}. But, to work with these functions would prove to be very awkward. Rather than the g_{ik}, it would be preferable to work with a complex which vanishes together with the M^{ik}, in order that the *vacuum* be realized if $M^{ik}=0$ and that then the variety automatically becomes Minkowskian if matter vanishes completely.

It is known today that the once contracted tensor $R^{kl} = R^{ikl}{}_i$ obtained from the Riemann-Christoffel tensor $R^{ikl}{}_m$ [2] has the property that it is necessary and sufficient to put it identically equal to zero in order to imply the (pseudo-)Euclidicity of the Riemannian variety. So we have to expect a relation which can be symbolically written as (R): $\underline{R} = \underline{R}(\underline{M})$ (where the underlining is a symbol of a tensor), built in a way such that $(M^{ik}=0) \to (R^{ik}=0)$.

Before getting there, let us notice that the M^{ik} are densities, in the physical sense of magnitudes referred to the unit of three-dimensional volume. In particular, in a local Galilean coordinate system whose time-

[1] For the notations used in Electrodynamics, see A. Mercier, Leçons sur les Principes de l'Electrodynamique classique (Neuchâtel 1952).

[2] For detailed definitions, see further below.

axis is in the direction of dx^0, M^{00} is, up to a factor c^2 the mass density ϱ as is observed by an observer who is bound with that system.

Now in Newtonian mechanics, e.g. the equation $\Delta\varphi = -4\pi G\mu$ applies. If we require Newtonian mechanics of gravitation to be an approximation of GRG, we have to expect (R) to be the generalization of the above equation. Therefore, $\underline{R}$ is the generalization of φ. Hence, it is $\underline{R}$ which is a "field". But instead of being a field of gravitation, it is a field of geometry and consequently g_{ik} too, as well as any magnitude built up by means of the g_{ik} 's and their derivatives as used in the construction of $\underline{R}$.

Furthermore, since every material point is identical with its worldline (just as was the case in special relativity), relation (R) will suffice to determine these lines. But these lines are the four-dimensional picture of trajectories, and trajectories are known from equations of motion. Hence, equations of motion are superfluous or redundant and they reduce to a mere interpretation of the kinematico-material situation. As long as one has to deal with material points which are electrically neutral, the principle of equivalence allows in the infinitesimal environment of every point $\boldsymbol{y}$ to look for a system of coordinates x^i in which no forces act and in reference to which the particle travels "along a straight line" i.e., following the greatest slope or, in other words, in the whole of its trajectory along the geodesic. But the geodesics are given by $\delta\int ds = 0$ and they are independent of the coordinate system. Therefore, worldlines coincide with geodesics and the equation of motion of a neutral particle whose only property is to bear mass will be the (differential) equation of the geodesic.

So, the whole problem of GRG is subsumed under the following: to determine $\underline{R}(\underline{M})$; in other words, *the universe would be a variety of Riemannian space–provided it is filled with nothing but masses*, of which it is no more necessary to say that they are grave. As a matter of fact, any energy distributed in that space will appear equivalent with inertial mass, hence with grave mass too. But this energy might have its origin in an interaction different from gravitation, in particular from electromagnetism or even, as has been known several years, from one of the weak or strong interactions. Then $\underline{R}(\underline{M})$ would be no more sufficient to describe the total state of the universe and a world-line will no more necessarily be a geodesic of $\mathbb{R}^4$. Moreover, it is not even certain that the description of the phenomena involved will succed in $\mathbb{R}^4$.

At the time when Einstein elaborated his theory of GRG and during quite a time afterwards, gravitation and electromagnetism alone were known. Atomic structure, even if it is of a quantum nature, does not require more than electromagnetism. People assumed at the beginning that it would be enough to add a tensor $\underline{T}$ of electromagnetic energy-momentum to the tensor $\underline{M}$. Surely, this is necessary in order to take the whole existent "mass" into account. But this does not explain electromagnetism which finds its expression, as is well known, in a Maxwellian theory which is alien to an Einsteinian relation

$$\underline{R} = R(\underline{M} + \underline{T})$$

and which consequently does not follow from it.

So, if on the one hand, the theory of special relativity was a theory of pure electromagnetism (without gravitation), GRG theory is on the other hand, a theory of gravitation without actually including electromagnetism. This is likely to be a very weak point since GRG theories have the pretention of being universal.

1.2. The Main Problems of GRG

A first problem arises, perhaps the most important: can GRG still be "generalized" once more in order to *include electromagnetism* in an intrinsic manner? Since this is the biggest problem, it may be the most difficult one too. In our days, physicists are inclined to believe that it cannot be solved. Whatever the answer is, one should notice before undertaking its analysis that a first formal difference immediately arises between gravitation and electromagnetism viz. the fact that the former one is of a fundamentally symmetric and polar nature (there are only material points of positive mass)[1] whereas the latter one is of a fundamentally anti-symmetric nature, i.e. it is solenoidal or dipolar according to the distinction between magnetism and electricity. In any case one should not be misled by a formal analogy between the formula of Coulomb and that of Newton; the former one is not valid unless restricted to electrostatics; so it must be ignored, for there is no electrostatics in the frame of Lorentz transformations. The metric tensor g_{ik} is symmetric: $g_{ik} = g_{ki}$. The same applies to R^{ik} and to M^{ik} (as well as to T^{ik}, which is the only symmetric character of electromagnetism!). From special relativity, we know that the tensor of the electromagnetic field F^{ik} is antisymmetric:

[1] In this argument we deliberately ignore the possibility that tachyons exist.

$F^{ik} = -F^{ki}$. It is always possible to construct symmetric tensors by means of anti-symmetric ones in products and combinations of products as is done for T^{ik}. The opposite is not feasable; one cannot obtain anti-symmetric tensors but by forming differences like $\frac{1}{2}(a^{ik} - a^{ki})$ as is done e.g. in a rotation; however, by taking symmetric tensors, these differences always vanish. So, one would have to start with a non-symmetric metric tensor, which is impossible within a Riemannian geometry. Therefore, the Riemannian frame work will have to be dropped in case electromagnetism shall be included otherwise than by admitting an independent Maxwellian theory to be, so to say glued onto GRG.

Moreover, another formal difference arises between GRG and electrodynamics: the equations for the Maxwellian field are linear, those for the Einsteinian field are not. This implies that in electrodynamics one can start from a superposition principle which is one of the conditions of quantization. On the contrary, it cannot be done, in principle and in all generality, within GRG. The quantization of the Einsteinian field of gravitation is therefore an obscure, if not a doubtful undertaking.

Many other problems arise: Beside the obscurity which has just been noticed, we can even ask *whether GRG and quantum theory are compatible* or whether they are in conflict. GRG ist not a quantized theory. In our days, quantization means field quantization and not simply the use of a Schroedinger equation. (Formally a Newtonian potential can be introduced in a Schroedinger equation, though physically this does not make much sense.) The next questions are than of a double nature.

On the one hand, since GRG postulates a generalization of the special relativistic variance to a covariance with respect to all changes of curvilinear coordinates, the theory of quantized fields, ordinarily restricted to Lorentz transformations should admit of a generally covariant form. In particular, the commutators, the distribution functions of the delta-type and other expressions like propagators will have to be put in that so-called generally covariant form. It is not at this place that a possible conflict arises and this problem is often considered as an academic affair without much practical value by non-relativists, since gravitation is known to be 10^{36} times weaker than the electromagnetic interaction.[1] (See below 7.2.2.)

[1] See e.g. H. J. Treder, Zur physikalischen Bedeutung einer Quantlung der Gravitation. Sitzungsber. der AdW der DDR, 3N, Akademie-Verlag, Berlin 1977, S. 29–45.

But on the other hand, on the view of the success of the quantization of the various fields which have been used so far to define fundamental particles, it is quite meaningful to ask whether one should quantize the gravitational field by some method which has still to be invented.[1] This suggests that there may be fundamental particles of the gravitational field: gravitons, just as there are the photons of the electromagnetic field, the mesons of the various mesonic fields etc. Now such a field is in principle always defined in its specificity which distinguishes it from another field by a Lagrangian density of a particular form. (He who talks Lagrangian implies a variation principle and hence the differential equations which follow from it.) There does exist a Lagrangian in GRG but its form is completely different from the Lagrangians of the ordinary quantifiable fields owing to the non-linearity of Einstein's equations. The ordinary quantifiable Lagrangians always show an oscillator-like structure, one charakter of which consists precisely in yielding linear equations. Gravitation has not the aspect of a field of oscillators. Maybe it has, in some approximation, or in a particular case but this would not satisfy rigorous requirements. So, on that point there might very well be a conflict or even an imcompatibility. More or less in the same connection another problem arises: are there *gravitational waves*? Furthermore, if the answer is yes, do they imply a *radiation*, i.e. the transport of some sort of energy? Originally, photons were the dual aspect of electromagnetic waves, for which the famous equation reads, for each component f of the field:

$$\Box f = 0 .$$

This equation is invariant during Lorentz transformations. Its generally covariant form will have to be established and one will have to find out if the components of the gravitational field are solutions of a similar equation under conditions analogous to the electromagnetic case (grossly speaking: in "vacuum"). In Newtonian mechanics, there was no question of waves because gravitation there was considered as an *immediate* action at distance, and there was no propagation with finite velocity. In GRG there is strictly speaking nothing but a network of lines, *world lines*, the spatio-temporal aspect of which–meaning: the spacial aspect and the temporal aspect separately–can only be noticed by local and momentary observations with the help of instruments bound to a Galilean coordinate

[1] The attempts made so far will be dealt with further on.

system in which the clock is explicited, i.e. in other words in the tangent Minkowskian space which is not identical with the variety, itself a model of the universe.

In vacuum, there is "nothing"; in it the M^{ik}'s vanish. But then if there are waves, what do they transport if the density of energy-momentum does not exist? Are they real signals? The existence of waves can be mathematically proved only under important restrictions and relativists do not agree among themselves as to the existence of gravitation radiation or signalization ("news") in GRG.

Another problem which arises and which is not alien to local and momentary measure or observation is that of the conservation of various magnitudes. The question will have to be cleared up: what is meant by conservation? We shall see that there again one deals with an essentially Galilean notion.

Apart from the problems mentioned so far which require *physical* interpretation, there are problems of *mathematical* nature. Differential equations always call for integration. Their *integration* implies two problems: the problem of initial conditions and that of evolution. On the whole it is Cauchy's problem.

However, the problem of integrability is distinct from that of finding particular solutions even if these solutions are rigorous or, as they are called, exact (non approximate). Several exact solutions can be classified according to certain properties, be they properties of symmetry, topology, stability or others.

Beside these problems which so to say arose from exterior and often already classic ways of tackling questions, GRG instigated original ones. The hugest one is doubtlessly the new *cosmology* implied by GRG. Cosmology is of course a very old field of investigation having its own classical solutions. But, what is peculiar to GRG is the analysis of the whole universe which it is supposed to accomplish: GRG is *cosmology itself*, if you wish, or at least it means to be. For, the only true procedure of GRG should consist in giving the expression of $\underline{R}$ in the whole universe, from which everything would follow in virtue of a rigorous and implacable determinism. Of course this is subject to many contingences. A first objection to that claim is that nature offers other phenomena than gravitation to our observation; GRG does not deal with such phenomena otherwise than by including in their gravitational form the energies attached to these phenomena but without treating the interactions responsible for them. The second objection is that one knows neither the distribution of energy-momentum in the universe, nor the values of the g_{ik}'s everywhere, which would be the same and would solve the problem. One can at most report about incomplete observations. So, one is led to make some hypothesis about the distribution, and to approach that what the world looks like. Another way out of the dilemma is to satisfy oneself with local descriptions, but that is no more true GRG cosmology. The latter thus proposes *cosmological models* which can be classified according to various characters. By the way, there are concurrent cosmologies, either by replacing GRG by a different mechanics, or by trying to generalize it once more. No doubt, each model will have to be submitted to astronomical observation, but the success of a decision is not yet at hand because observations on the cosmological scale are few. The most manifest phenomena it has revealed and about which everybody has heard is the red shift of the lines of the spectrum from far distant matter called Hubble effect, which is usually identified with a Doppler effect, but which is possibly due to another cause than the increase in time of the distance of the sources.

This gives us the opportunity to strife the problem of the experimental verification of the theory. Contrarily to a common opinion, we shall be led to say after due consideration that the verification of GRG is highly satisfactory. However, with regard to what is wrongly called crucial effects, these are small and their observation is not easy. By the way, when talking about experimental verifications, one should realize that the theory of GRG was not developed for experimental reasons, not even for purely theoretical reasons, but for reasons the very nature of which is epistemological. By the way, already the theory of special relativity had been elaborated for theoretical as well as for experimental motives. Indeed and contrarily to what is usually said, not only the negative result of Michelson's experiment led Einstein to its formulation but also the discrepancy which was found between the variance of Newtonian mechanics and the variance of electrodynamics. In the search for a unification of mechanics and electrodynamics, gravitation had to be dropped from the descriptions allowed by special relativity. For epistemological reasons, gravitation had to be reintroduced. Was that the only reason for generalizing special relativity? In a way, yes. But another one can be invoked, namely that there is no sufficient reason to restrict oneself to Lorentz transformations in any argument about the invariance of fundamental equations. This application of the principle of sufficient reason has, by the way, worked as a sensation and convinced many people. But the argument can be redoubled by saying that there is no sufficient reason to restrict oneself to the use of a Riemannian variety and so on, so the series of generalizations has no end. On going back to the generalization realized in GRG, it was found that electromagnetism had in its turn to be, so to say, forgotten, if not dropped altogether. This, in a way marks the second step of a zig-zag search for a *unified theory* (in French "unitaire", in German "einheitlich" where these three adjectives have a slightly different meaning according to the vocabulary) of gravitation and electromagnetism.

Finally, two problems remain. The first one concerns "mecano-dynamics" of GRG, the second "geometro-dynamics".

By *mecano-dynamics* we mean the following. We know that the idea of general covariance allows to go from one system of coordinates to another without influencing the intrinsic physical situation. This is why one should distinguish that which in the theory is really dynamical from that which concerns coordinates, in order to make sure that by changing coordinates, dynamics remains unchanged. This problem is connected with both gauge and the canonical formalism. And since canonical formalism is one of the densest forms of mechanics, we talk here of "mecano-dynamics".

Geometro-dynamics is another aspect of dynamics consisting in the fact that according to GRG one is confronted with gravitation without it being necessary to introduce a particular phenomenon called gravitation. As Wheeler puts it: it is gravitation without gravitation, since there is nothing but geometry. Fields, particles, physical objects are according to this interpretation of GRG mere constructs from geometry. This, as far as one claims to describe a so-called dynamical behaviour, this description will never be anything else than a "geometro-dynamics". It is an illusion to believe that there is matter.

1.3. Difficulties of the GRG Theory

These difficulties can be grouped according to three main view points. Firstly, GRG theory claims to be universal but is not: hence, it must be submitted to further generalizations or unifications which never get to an end. Then, it is in two respects opposed to the idea of a quantum theory; on the one hand, it is deterministic in a particularly rigorous sense which is irreducible to the quantum interpretation of a probabilistic determinism;

on the other hand, its field equations are non-linear, which renders any attempt of a quantization more than difficult. Thirdly, as far as it goes beyond what is already known from the Newtonian theory of gravitation, GRG is experimentally verifiable only by means of very weak effects, since all the stronger effects were already explained by pre-relativistic mechanics. This last remark explains why GRG has been held during years in a kind of disgrace by a great many physicists, not because they thought it to be wrong, but because they found it inefficient, considering the enormous mathematical undertakings it required.

Moreover, just as special relativity itself, GRG has an epistemological aspect in disagreement with the strictly Kantian point of view of natural philosophy. For this reason, it has a number of antagonists not only among professional philosophers but also among an audience unconscious of how strong they are attached to a teaching imbued of the Kantism which they have received at the secondary school level.

In a similar connection GRG is a kind of universal cosmology *à la Descartes* which easily yields a pantheistic view of the Creation; this has set its author into a philosophical tradition going back to Spinoza. It has also made of GRG the source of religious if not theological difficulties. Indeed, from times immemorial, cosmology has been either one of the grounds on which religions arose, or an intellectual product of the reflexion of those who have exposed and explained the doctrines of their own religion. Since Newtonian mechanics was known and still with an impregnable authority since the theory of GRG, the human mind stands face to face with a cosmology whose agreement with reality surpasses everything a religion has ever been able to propound.

Finally, there are a number of difficulties inherent to GRG:

– the impossibility to establish a theorem of conservation of energy in the usual sense,

– the quasi-impossibility to strictly dispose of a test particle,

– the difficulty to trace space and time in a way simultaneously adapted to the measurements made by means of the apparatuses in use and accessible to the intuition the human mind has of the prime-notions and of the present conditions,

– the fact that Riemannian geometry is a local geometry and not a global one, preventing from encompassing the whole universe by means of GRG, that is, in its classical form, implying the persistence of the critique uttered by Mach about physics altogether,

– the inexistence of any rigidity of matter,
– and even the impossibility to give preference, in principle, to any coordinate system susceptible of a natural realization in bodies either terrestrial or celestial.

For all these reasons, the common physicist does not feel particularly attracted towards this domain of research and thought. It requires a certain taste for abstraction and for philosophical speculation. Recently however astrophysical discoveries have revealed that there happen throughout the universe strange facts, the explanation of which will most probably request a theory at least as powerful and original as GRG. And, since we possess the latter, its value in the stock exchange has considerably gone up, if we may say so.

2. FORMALISM OF FOUR-DIMENSIONAL VARIETIES

2.1. Differentiable Manifolds

We shall start with four real variables y_k $(k=1,2,3,4)$, called coordinates, susceptible of taking each independently all the values between given limits. Let y be the set of the four coordinates. Let $\varphi(y)$ be a function of the coordinates.

We shall write

$$\varphi_{,k} \equiv \frac{\partial \varphi}{\partial y^k}$$

for the derivative of φ with regard to y^k, $\varphi_{,k,l}$ etc. for the derivatives of higher order. All the functions which we shall consider will be assumed to possess at least derivatives of the orders one and two presenting discontinuities exceptionally, namely through certain surfaces. These functions are called functions of class C^2 by bits.

From the coordinates y, one can define new coordinates y' by giving four functions

$$y'^l = \varphi'^l(y) .$$

We shall require the functions φ'^l to be of class C^2 by bits with non-vanishing Jacobian.

The ensemble of points y or y', ... constitute a differentiable manifold where changes of coordinates are reversible.

Choosing the number of coordinates equal to 4 is done with the purpose to apply the formalism to GRG.

A network of four equations

$$F_l(y) = 0$$

defines a point or event at the crossing of 4 hyper-surfaces, e.g. fixed values of the y^k. Let $\boldsymbol{M}$ be such a point from which 4 lines diverge along the intersections of the 4 hyper-surfaces. These lines can be chosen as the coordinate lines themselves by keeping 3 of the coordinates constant and varying the fourth one. Then $\frac{\mathrm{d}y^k}{|\mathrm{d}y^k|}$ is a vector tangent to the line y^k, and by making $k=1, 2, 3, 4$ one obtains a *tetrapod* or *tetrad* of tangent unit vectors.

At the point $\boldsymbol{M}+\mathrm{d}\boldsymbol{M}$ defined by $y^k+\mathrm{d}y^k$, another tetrapod will be found and so on for all points of the manifold.

In each of these points, the tetrapod can be used as a Cartesian coordinate system within an infinitesimal neighbourhood. This system will be called *natural*.

2.2. Metric Space $\mathbb{M}^4$

Space is metric if the square of a distance between $\boldsymbol{M}$ and $\boldsymbol{M}+\mathrm{d}\boldsymbol{M}$ is defined by

$$\mathrm{d}s^2 = g_{ik}\mathrm{d}y^i\mathrm{d}y^k, \quad g_{ik} = g_{ik}(y)$$

where $g_{ik}(y)$ is a function of class C^2 by bits.

We shall assume $\mathbb{M}^4$ to be of *normal hyperbolic type*, i.e. endowed with a metric such that in the infinitesimal neighbourhood of each point a change of coordinates $y \to x$ can be found leading to a $\mathrm{d}s^2$ of the simple form

$$\mathrm{d}s^2 = (\mathrm{d}x^0)^2 - [(\mathrm{d}x^1)^2 + (\mathrm{d}x^2)^2 + (\mathrm{d}x^3)^2] \,.$$

The corresponding g_{ik}, then called $\overset{\circ}{g}_{ik}$, assume the form

$$(\overset{\circ}{g}_{ik}) = \begin{pmatrix} 1 & 0 & 0 & 0 \\ 0 & -1 & 0 & 0 \\ 0 & 0 & -1 & 0 \\ 0 & 0 & 0 & -1 \end{pmatrix} .$$

The corresponding natural coordinate system is then orthogonal and the neighbourhood is of pseudo-Euclidean type. Space thus defined is called *space-time*. This variety is *Riemannian* if

$$g_{ik} = g_{ki}$$

(symmetry condition). It will be designated by $\mathbb{R}^4$. A pseudo-Euclidean variety will be designated by $\mathbb{E}^4$.

In the neighbourhood of a point $\boldsymbol{M}$, $\mathrm{d}s^2$ can be put equal to zero:

$$\mathrm{d}s^2 = 0$$

without all the $\mathrm{d}x^i$ vanishing. Hence, in each point, there is an *elementary* isotropic *cone*. The $\mathrm{d}x^i$ are linearly independent local Pfaffian forms. It is always possible to go back to the y^k from the x^i. The $\mathrm{d}x^i$ build a vector, $\mathrm{d}\boldsymbol{x}$, the same, by the way, up to higher infinitesimal orders, as $\mathrm{d}\boldsymbol{M}$ or $\mathrm{d}\boldsymbol{y}$. Given a $\mathrm{d}\boldsymbol{x}$ (that is given $\mathrm{d}x^i$, hence also $\mathrm{d}y^k$) the vector $\mathrm{d}\boldsymbol{x}$ as well as $\mathrm{d}\boldsymbol{M}$ are called

time-like if	$g_{ik}\mathrm{d}x^i\mathrm{d}x^k = g_{ik}\mathrm{d}y^i\mathrm{d}y^k$		> 0
isotropic if	”	”	$= 0$
space-like if	”	”	< 0 .

By giving a quadruple of functions

$$y^k = \varphi^k(\sigma)$$

of a continuous parameter σ, e.g. $\sigma = s$ or $\sigma = \Psi(s)$, one defines a line Γ. At a point where $\mathrm{d}\boldsymbol{y}$ or $\mathrm{d}\boldsymbol{M}$ are time-like along Γ (respectively space-like), Γ is time-like (space-like) too. Mathematically, the orientation (time-like, isotropic, space-like) can vary along one and the same line. But if a line is to have a physical meaning, the question may arise whether it is physically possible that its orientation varies from point to point.

The space $\mathbb{E}^4$ tangent to $\mathbb{R}^4$ is a Minkowskian variety, the structure of which is defined by the kinematical axioms of special relativity.

2.3. Contravariance and Covariance

A set of four vectors $\boldsymbol{e}_i$ such that

$$\mathrm{d}\boldsymbol{M} = \boldsymbol{e}_i \mathrm{d}y^i$$

is called a *basis*. By a change of coordinates $y \to y'$,

$$\mathrm{d}y'^i = a'^i_k \mathrm{d}y^k \quad \left(a'^i_k \equiv \frac{\partial y'^i}{\partial y^k}\right),$$

$$\mathrm{d}y^k = a^k_i \mathrm{d}y'^i \quad \left(a^k_i = \frac{\partial y^k}{\partial y'^i}\right).$$

These relations define a pattern of *contravariance*: each so-called tensorial magnitude having an upper index k and transforming like $\mathrm{d}y^k$ with regard to that index, is called contravariant for that index.

After a change of coordinates, we get a new basis e_l for which

$$e'_l \mathrm{d}y'^l = e_k \mathrm{d}y^k$$

or

$$e'_i \frac{\partial y'^i}{\partial y^m} \mathrm{d}y^m = e_k \frac{\partial y^k}{\partial y'^l} \mathrm{d}y'^l,$$

hence

$$e'_l = e_k \frac{\partial y^k}{\partial y'^l} = e_k a^k_l$$

$$e_k = e'_i \frac{\partial y'^i}{\partial y^k} = e'_i a'^i_k.$$

These relations define the pattern of *covariance*.

In general a tensor $A^{\alpha\beta\ldots}_{\lambda\mu\ldots}$ is said to be covariant with regard to $\lambda\mu \ldots$ and contravariant with regard to $\alpha\beta \ldots$ if it transforms as follows under coordinate changes

$$A'^{ij\ldots}_{pq\ldots} = a'^i_\alpha a'^j_\beta \ldots\ldots a^\lambda_p a^\mu_q \ldots\ldots A^{\alpha\beta\ldots}_{\lambda\mu\ldots}.$$

As an example, the metric g_{ik} is a twice covariant tensor, which warrants the *invariance* of $\mathrm{d}s^2$.

Invariants $I' = I$ under coordinate changes are also called scalars (tensors of zero order); they are particularly interesting.[1]

The existence of a metric is not necessary for the definition of tensors which already exist in a simple affine space.

[1] Some authors distinguish between (mathematical) *invariants* and (physical) *scalars* (from "scale").

2.4. Affine Connection

Independently of the imposition of a metric, we can define the *affine connection*. One starts from a manifold of points $\boldsymbol{M}$, $\boldsymbol{M}+\mathrm{d}\boldsymbol{M}$ . . . defined by values y^i, $y^i+\mathrm{d}y^i$ . . . of the coordinates. Let $\boldsymbol{e}_k$ (or $\boldsymbol{e}_i'$) be a basis in $\boldsymbol{M}$ such that $\mathrm{d}\boldsymbol{M}=\boldsymbol{e}_i\mathrm{d}y^i$. In $\boldsymbol{M}+\mathrm{d}\boldsymbol{M}$, it will be $\boldsymbol{e}_k+\mathrm{d}\boldsymbol{e}_k$. Let the coordinates be changed $y\to y'$. The question arises: how is the change of coordinates to be made in $\boldsymbol{M}+\mathrm{d}\boldsymbol{M}$ if it is known in $\boldsymbol{M}$?

It must be possible to put

$$\mathrm{d}\dot{\boldsymbol{M}}=\boldsymbol{e}_i\omega^i=\boldsymbol{e}_i\mathrm{d}y^i\,,$$
$$\mathrm{d}\boldsymbol{e}_k=\boldsymbol{e}_i\omega_k^i\,.$$

One defines an entity Γ_{kl}^i such that $\mathrm{d}\boldsymbol{e}_k$ be linear in the $\mathrm{d}y^i$ (just as $\mathrm{d}\boldsymbol{M}$):

$$\mathrm{d}\boldsymbol{M}=\boldsymbol{e}_i\mathrm{d}y^i\,,$$
$$\mathrm{d}\boldsymbol{e}_k=\Gamma_{kl}^i\boldsymbol{e}_i\mathrm{d}y^l\,.$$

These Γ_{kl}^i allow to *connect* the tetrapods from a point to a neighbouring point. In order to define an *affine* space the Γ_{kl}^i must be given. For this reason these Γ_{kl}^i are called the *affine connection*. Affine space can be constructed point by point by giving the affine connexion as a function of y.

2.5. Integrability, Torsion, Curvature

Going back to metric space $\mathbb{M}^4$ defined by $g_{ik}(y)$ we shall postulate the symmetry of the g_{ik}.

Consider the natural coordinate system in a point $\boldsymbol{M}$ of $\mathbb{M}^4$ by prolonging its axes but by keeping constant the values of the g_{ik} which will then to be called $(g_{ik})_0$. We get Cartesian axes of coordinates x^i which define a space $\mathbb{M}_0^4$ where the square of the distance is everywhere

$$\mathrm{d}\sigma^2=(g_{ik})_0\,\mathrm{d}x^i\mathrm{d}x^k\,.$$

This can also be written with the help of a constant basis $(\boldsymbol{e}_i)_0$ equals to $\boldsymbol{e}_i$ in M:

$$\mathrm{d}\sigma^2=(\boldsymbol{e}_i)_0\,(\boldsymbol{e}_k)_0\,\mathrm{d}x^i\mathrm{d}x^k,\quad (g_{ik})_0=(\boldsymbol{e}_i)_0\cdot(\boldsymbol{e}_k)_0\,.$$

In each point of M^4 the same applies. Up to first order magnitudes, relations valid in M^4 are the same as in M_0^4:

$$\mathrm{d}\boldsymbol{M}=(\mathrm{d}\boldsymbol{M})_0 \rightarrow \left(\frac{\partial M}{\partial x^k}\right)_0 = \frac{\partial M}{\partial y^k} .$$

However what is inequal is

$$(\mathrm{d}\boldsymbol{e}_k)_0 \neq \mathrm{d}(\boldsymbol{e}_k)_0 ,$$

for

$$(\mathrm{d}\boldsymbol{e}_k)_0=(\Gamma^i_{kl})_0\,(\boldsymbol{e}_i)_0\,(\mathrm{d}y^l)_0=(\Gamma^i_{kl})_0\,(\boldsymbol{e}_i)_0\,\mathrm{d}x^l \simeq \Gamma^i_{kl}\boldsymbol{e}_i\mathrm{d}x^l ,$$
$$\mathrm{d}(\boldsymbol{e}_k)_0=\Gamma^i_{kl}\boldsymbol{e}_i\mathrm{d}y^l \quad (\mathrm{d}y^l \neq \mathrm{d}x^l!) .$$

If one wants to postulate the equality of these two magnitudes: $(\mathrm{d}\boldsymbol{e}_k)_0=\mathrm{d}(\boldsymbol{e}_k)_0$, then

$$(\boldsymbol{e}_l)_0\,(\mathrm{d}\boldsymbol{e}_k)_0+(\boldsymbol{e}_k)_0\,(\mathrm{d}\boldsymbol{e}_l)_0=(\boldsymbol{e}_l)_0\,\mathrm{d}(\boldsymbol{e}_k)_0+(\boldsymbol{e}_k)_0\,\mathrm{d}(\boldsymbol{e}_l)_0 ,$$

hence

$$(\mathrm{d}g_{ik})_0=(\mathrm{d}\,(\boldsymbol{e}\cdot\boldsymbol{e}_k))_0=(\boldsymbol{e}_l)_0\,(\mathrm{d}\boldsymbol{e}_k)_0+(\boldsymbol{e}_k)_0\,(\mathrm{d}\boldsymbol{e}_l)_0$$
$$=(\boldsymbol{e}_l)_0\,\mathrm{d}(\boldsymbol{e}_k)_0+(\boldsymbol{e}_k)_0\,\mathrm{d}(\boldsymbol{e}_l)_0=\mathrm{d}(g_{ik})_0 .$$

Else

$$\boldsymbol{e}_i=(\boldsymbol{e}_i)_0+(\Gamma^k_{il})_0\,(y^l-x^l)\,(\boldsymbol{e}_k)_0 \quad \text{(where } (y^l-x^l) \text{ plays the role of } \mathrm{d}y)$$

$$(\mathrm{d}\boldsymbol{M})_0=(\boldsymbol{e}_i\mathrm{d}y^i)_0=((\boldsymbol{e}_i)_0\,\mathrm{d}y^i+(\Gamma^k_{il})_0\,(y^l-x^l)\,\mathrm{d}y^i(\boldsymbol{e}_k)_0)_0$$
$$=(\boldsymbol{e}_k)_0\,[\mathrm{d}y^k+(\Gamma^k_{il})_0\,(y^l-x^l)\,\mathrm{d}y^i]_0 ,$$

which shows us that, if $i \neq l$

$$\frac{\partial \boldsymbol{M}}{\partial y^i}=(\boldsymbol{e}_i)_0+(\boldsymbol{e}_k)_0\,(\Gamma^k_{il})_0\,(y^l-x^l), \quad \frac{\partial \boldsymbol{M}}{\partial y^l}=(\boldsymbol{e}_l)_0+(\boldsymbol{e}_k)_0\,(\Gamma^k_{li})_0\,(y^i-x^i)$$

$$\left(\frac{\partial^2 \boldsymbol{M}}{\partial y^l \partial y^i}\right)_0=(\boldsymbol{e}_k)_0\,(\Gamma^k_{il})_0, \qquad \left(\frac{\partial^2 \boldsymbol{M}}{\partial y^i \partial y^l}\right)_0=(\boldsymbol{e}_k)_0\,(\Gamma^k_{li})_0 .$$

Therefore in general,

$$\left(\frac{\partial^2 \boldsymbol{M}}{\partial y^l \partial y^i}\right)_0 \neq \left(\frac{\partial^2 \boldsymbol{M}}{\partial y^i \partial y^l}\right)_0 \quad \text{in} \quad \mathsf{M}_0^4 ,$$

from which follows the non-integrability of a line

$$\oint \mathrm{d}\boldsymbol{M} \neq 0 ,$$

unless

$$\Gamma^k_{il} = \Gamma^k_{li} .$$

The best way to define a *Riemannian variety* $\mathbb{R}^4$ is to assume the integrability of lines in $\mathbb{M}^4$. Thus we see that a necessary and sufficient condition for a variety to be Riemannian consists in the symmetry between the two lower indices of the affine connection.

In $\mathbb{M}^4$ a *torsion* is defined by the (contravariant) vector

$$\Omega^i = \frac{1}{2} (\Gamma^i_{jk} - \Gamma^i_{kj}) (\mathrm{d}y^k \delta y^j - \mathrm{d}y^j \delta y^k)$$

where d and δ are taken along the adjacent sides of an infinitesimal parallelogram whose area yields precisely projections

$$\frac{1}{2} (\mathrm{d}y^k \delta y^j - \mathrm{d}y^j \delta y^k) .$$

Hence, a variety $\mathbb{M}^4$ has in general a torsion, whereas a Riemannian one has none.

In $\mathbb{M}^4$ a *curvature* is defined by the tensor

$$\Omega^l_k = R^l_{kpq} \cdot \frac{1}{2} (\mathrm{d}y^q \delta y^p - \mathrm{d}y^p \delta y^q)$$

where R^l_{kpq}, called the Riemann-Christoffel tensor, is defined by:

$$R^l_{kpq} \equiv \Gamma^l_{kp,q} - \Gamma^l_{kq,p} + \Gamma^s_{kp}\Gamma^l_{sq} - \Gamma^s_{kq}\Gamma^l_{sp} , \quad \Gamma^c_{ab,j} \equiv \frac{\partial \Gamma^c_{ab}}{\partial y^j} .$$

In the tangent space where the $(g_{ik})_0$ are constant, the Γ's clearly vanish, since the $\boldsymbol{e}_i$ are also constant. Therefore, tangent space possesses neither curvature nor torsion; it is Euclidean (or pseudo-Euclidean according to the signature). In it there is no objection to write

$$g_{ik} = g_{ki} , \quad \text{for,} \quad \text{in}$$

$$\mathrm{d}\sigma^2 = (g_{12})_0 \, \mathrm{d}x^1 \mathrm{d}x^2 + \ldots + (g_{21})_0 \, \mathrm{d}x^2 \mathrm{d}x^1 + \ldots ,$$

the two terms

$$\ldots (g_{12}+g_{21})_0\, dx^1 dx^2 + \ldots$$

cannot be distinguished. Assuming symmetry is thus not a restriction.

Reciprocally, it can be proved that

1° if $R^l_{kpq}=0$ everywhere and for all indices, space is Euclidean.

2° if $R^l_{kpq} \neq 0$ but $\Omega^i=0$, space is Riemannian.

2.6. Passage from Covariance to Contravariance and Reciprocally in a Metric Space

Let e_i be the basis.

We can write $d\boldsymbol{y}=\boldsymbol{e}_i dy^i=\boldsymbol{e}_k dy^k$,

hence

$$ds^2=d\boldsymbol{y}\cdot d\boldsymbol{y}=\boldsymbol{e}_i\cdot\boldsymbol{e}_k dy^i dy^k.$$

Thus in a metric space,

$$g_{ik}=\boldsymbol{e}_i\cdot\boldsymbol{e}_k.$$

Build the determinant

$$g\equiv\|g_{ik}\|$$

as well as its minors γ^{ik} such that $g_{ik}\gamma^{ik}=g$. The symbols

$$g^{ik}=\frac{\gamma^{ik}}{g}$$

are contravariant, since

$$g_{ik}g^{ik}=1\ ,$$

an invariant which is only possible if g^{ik} is itself contravariant, since g_{ik} is covariant.

From the properties of determinants, we know that,

$$g_{ik}g^{kl}=\delta^l_i\ .$$

Moreover, let $\boldsymbol{A}=A^k\boldsymbol{e}_k$ be a vector of contravariant components A^k. Its covariant components will be:

$$A_k=\boldsymbol{A}\cdot\boldsymbol{e}_k=A^l\boldsymbol{e}_l\cdot\boldsymbol{e}_k=A^l g_{lk}\ .$$

Therefore g_{lk} allows to lower an index from contravariance to covariance. Reciprocally, by defining g^{lm} such that

$$g_{im}g^{kl}=\delta_i^l\,,$$

we get

$$g^{ik}A_k=g^{ik}A^l g_{lk}=\delta_l^i A^l=A^i\,.$$

Hence the g^{ik} allow to transform covariant indices into contravariant indices.

This can be generalized to any tensor:

$$g_{pm}A^{ik\ldots m\ldots}=A^{ik\ldots}{}_p{}^{\ldots}$$

$$g^{ip}A_{qr\ldots p\ldots}=A_{qr\ldots}{}^i{}_{\ldots}$$

Note that the position of the indices must be respected in the horizontal order.

2.7. Relations between the Affine Connection and the Metric in $\mathbb{R}^4$

We have

$$\mathrm{d}e_k=\Gamma^i_{kl}e_i\mathrm{d}y^l\,,$$

and

$$\mathrm{d}g_{kj}=\mathrm{d}\,(e_k\cdot e_j)=e_k\Gamma^i_{jl}\cdot e_i\mathrm{d}y^l+\Gamma^i_{kl}e_i\mathrm{d}y^l\cdot e_j=(g_{ki}\Gamma^i_{jl}+g_{ij}\Gamma^i_{kl})\,\mathrm{d}y^l$$

hence

$$\partial g_{kj}/\partial y^l=g_{jk,l}=g_{ik}\Gamma^i_{jl}+g_{ij}\Gamma^i_{kl}\,.$$

By combination, all calculations done, it follows that

$$\Gamma^i_{kl}g_{im}=\frac{1}{2}\,[g_{lm,k}+g_{km,l}-g_{kl,m}]\equiv[kl,\,m]\,.$$

$[kl, m]$ is called *Christoffel symbol of the first kind*. Then, in the same way

$$\Gamma^i_{kl}=\frac{1}{2}\,g^{im}\,(g_{lm,k}+g_{km,l}-g_{kl,m})\equiv\left\{{i \atop kl}\right\}.$$

$\left\{{i \atop kl}\right\}$ is called *Christoffel symbol of the second kind*.

This implies

$$\Gamma^i_{kl} = \begin{Bmatrix} i \\ kl \end{Bmatrix} \quad \text{in} \quad \mathbb{R}^4 .$$

In $\mathbb{R}^4$, thereafter,

$$\mathrm{d}\boldsymbol{e}_p = \begin{Bmatrix} i \\ pl \end{Bmatrix} \boldsymbol{e}_i \mathrm{d}y^l .$$

Einstein's Remark on Symmetry: the symmetry between the indices p and l in $\begin{Bmatrix} i \\ pl \end{Bmatrix}$ or Γ^i_{pl} has no counter-value in the relation

$$\mathrm{d}\boldsymbol{e}_p = \begin{Bmatrix} i \\ pl \end{Bmatrix} \boldsymbol{e}_i \, \mathrm{d}y^l$$

between $\mathrm{d}y^l$ and $\mathrm{d}\boldsymbol{e}_p$. There is no reason 'a priori' to choose a symmetric affine connection,–and a symmetric metric.

Let us still ascertain that g_{kl} and $\begin{Bmatrix} i \\ pq \end{Bmatrix}$ are simultaneously symmetric (or non-symmetric). It is clear first of all, that the symmetry of the g_{ik} involves the symmetry of Γ^i_{kl}, and that their non-symmetry involves non-symmetry of Γ^i_{kl}.

The reciprocal relation is true, because, as $\underline{\boldsymbol{M}}_{,i,k} = \underline{\boldsymbol{M}}_{,k,i}$ if the Γ's are symmetric,

$$\begin{aligned}
(\boldsymbol{M} \cdot \boldsymbol{M})_{,i} &= \boldsymbol{M}_{,i} \cdot \boldsymbol{M} + \boldsymbol{M} \cdot \boldsymbol{M}_{,i} , \\
(\boldsymbol{M} \cdot \boldsymbol{M})_{,i,k} &= \boldsymbol{M}_{,i,k} \cdot \boldsymbol{M} + \boldsymbol{M}_{,i} \cdot \boldsymbol{M}_{,k} + \boldsymbol{M}_{,k} \cdot \boldsymbol{M}_{,i} + \boldsymbol{M} \cdot \boldsymbol{M}_{,i,k} \\
&= \boldsymbol{M}_{,k,i} \cdot \boldsymbol{M} + \boldsymbol{M}_{,k} \cdot \boldsymbol{M}_{,i} + \boldsymbol{M}_{,k} \cdot \boldsymbol{M}_{,k} + \boldsymbol{M} \cdot \boldsymbol{M}_{,k,i} \\
&= (\boldsymbol{M} \cdot \boldsymbol{M})_{,k,i}
\end{aligned}$$

and if $\boldsymbol{M} \cdot \boldsymbol{M} = \mathrm{d}s^2$,

$$g_{ik} = \frac{\partial^2 s}{\partial y^i \partial y^k} = \frac{\partial^2 s}{\partial y^k \partial y^i} = g_{ki} .$$

2.8. Summary of the Classification of Metric Varieties

There are three successive classes of metric varieties.

$\mathbb{E}^4$ (Euclidean)	$\begin{Bmatrix}\oint dM=0 \\ \oint de_i=0\end{Bmatrix}$	neither torsion, nor curvature $g_{ik}=\text{const}=g_{ki}$, $\Gamma^i_{kl}=0$;
$\mathbb{R}^4$ (Riemannian)	$\begin{Bmatrix}\oint dM=0 \\ \oint de_i \neq 0\end{Bmatrix}$	curvature, but no torsion $g_{ik}=g_{ki}=g_{ik}(y)$, $\Gamma^i_{lk}=\Gamma^i_{kl}\neq 0$;
$\mathbb{M}^4$ (whatever)	$\begin{Bmatrix}\oint dM\neq 0 \\ \oint de_i \neq 0\end{Bmatrix}$	torsion and curvature $g_{ik}\neq g_{ki}\neq 0 \quad \Gamma^i_{kl}\neq\Gamma^i_{lk}$.

2.9. Absolute Differential and Covariant Derivation

Let us consider, in an affine space $\mathbb{A}^4$, a vector field

$$\boldsymbol{A}=A^i\boldsymbol{e}_i=\boldsymbol{A}(y)$$

and its variation $\mathrm{d}\boldsymbol{A}$ resulting from arbitrary $\mathrm{d}y^i$ variations:

$$\begin{aligned}\mathrm{d}\boldsymbol{A}&=\mathrm{d}A^i\boldsymbol{e}_i+A^i\mathrm{d}\boldsymbol{e}_i=\mathrm{d}A^i\boldsymbol{e}_i+A^k\mathrm{d}\boldsymbol{e}_k\\ &=(\mathrm{d}A^i+\omega^i_kA^k)\,\boldsymbol{e}_i,\quad \mathrm{d}\boldsymbol{e}_k=\omega^i_k\boldsymbol{e}_i=\Gamma^i_{kl}\mathrm{d}y^l\boldsymbol{e}_i\\ &\equiv DA^i\boldsymbol{e}_i\end{aligned}$$

where we have put

$$DA^i\equiv \mathrm{d}A^i+\Gamma^i_{kl}A^k\mathrm{d}y^l\ .$$

DA^i, called *absolute differential*, is a contravariant vector, since $DA^i\boldsymbol{e}_i=$ $=d\boldsymbol{A}$ is the invariant vector $d\boldsymbol{A}$ itself.

The absolute differential consists therefore of two increases, the first dA^i, owing to the variations of the components of A^i when passing from y to $y+\mathrm{d}y$, the second, owing to the modification of the natural coordinate system during the same passage.

We can then form a tensor

$$\nabla_l A^i\equiv A^i_{;l}\equiv\frac{DA^i}{\mathrm{d}y^l}=A^i_{,l}+\Gamma^i_{kl}A^k$$

called *covariant derivative* of A^i (contravariant) which can be written either $A^i_{;l}$ or $\nabla_l A^i$.

To the contravariant vector A^i corresponds a covariant vector such that A^iA_i is invariant; it is equal to $\boldsymbol{A}\cdot\boldsymbol{e}_i$. The corresponding increases $(\mathrm{d}\boldsymbol{A})_i=\mathrm{d}\boldsymbol{A}\cdot\boldsymbol{e}_i$ which we shall write down as DA_i, are equal to

$$\begin{aligned}DA_i&\equiv \mathrm{d}\boldsymbol{A}\cdot\boldsymbol{e}_i=\mathrm{d}(\boldsymbol{A}\cdot\boldsymbol{e}_i)-\boldsymbol{A}\cdot\mathrm{d}\boldsymbol{e}_i\\&=\mathrm{d}A_i-\boldsymbol{A}\cdot\Gamma^k_{il}\boldsymbol{e}_k\mathrm{d}y^l\\&=\mathrm{d}A_i-\Gamma^k_{il}A_k\mathrm{d}y^l.\end{aligned}$$

From this, the covariant derivative of A_i (covariant) is found as

$$A_{i;l}\equiv\nabla_lA_i=A_{il}-\Gamma^k_{il}A_k.$$

For a tensor of the $\left\{\begin{matrix}n.\text{ th covariant order}\\ m.\text{ th contravariant order}\end{matrix}\right\}$

$$\begin{aligned}A^{\cdots}_{\cdots;l}&\equiv\nabla_lA_{ik\ldots}{}^{pq\ldots}\\&\equiv A_{ik\ldots}{}^{pq\ldots}{}_{,l}+\Gamma^p_{sl}A_{ik\ldots}{}^{sp\ldots}+\Gamma^q_{sl}A_{ik\ldots}{}^{ps\ldots}+\ldots\\&\quad-\Gamma^s_{il}A_{sk\ldots}{}^{pq\ldots}-\Gamma^s_{kl}A_{is\ldots}{}^{pq\ldots}.\end{aligned}$$

In a *metric* space, the Γ's coincide with the { }. Furthermore, as can be verified,

$$DA_l=g_{ls}DA^s.$$

In a *Riemannian* space, it is particularly interesting to calculate

$$g_{ik;l}=g_{ik,l}-\Gamma^s_{il}g_{sk}-\Gamma^s_{kl}g_{is}.$$

For, in $\mathbb{R}^4$,

$$g_{ij}=g_{ji},\quad \Gamma^s_{il}=\Gamma^s_{li},\quad \Gamma\to\{\},\quad [ik,l]=g_{sl}\left\{\begin{matrix}s\\ik\end{matrix}\right\},$$

hence

$$\begin{aligned}g_{ik;l}&=g_{ik,l}-\Gamma^s_{il}g_{sk}-\Gamma^s_{kl}g_{si}\\&=g_{ik,l}-[il,k]-[kl,i]\\&=g_{ik,l}-\frac{1}{2}\left[g_{kl,i}+g_{ik,l}-g_{il,k}+g_{ki,l}+g_{li,k}-g_{lk,i}\right]\\&=0.\end{aligned}$$

The covariant derivative of the metric vanishes indentically in $\mathbb{R}^4$.

2.10. Parallel Displacement or Transport

There is a difference between an *arbitrary* differential and the *absolute* differential:

$$\mathrm{d}A^k = DA^k + (-\Gamma^k_{lm} A^l \mathrm{d}y^m)$$

or

$$\mathrm{d}A_k = DA_k + (+\Gamma^l_{km} A_l \mathrm{d}y^m) .$$

Hence, the arbitrary displacement is „more" than the absolute differential. This supplement is called

$$\text{parallel displacement} \begin{cases} (\mathrm{d}A^k)_\parallel \equiv -\Gamma^k_{lm} A^l \mathrm{d}y^m \\ (\mathrm{d}A_k)_\parallel \equiv +\Gamma^l_{km} A_l \mathrm{d}y^m. \end{cases}$$

Other interesting displacements will still be defined, namely of Fermi and of Walker (see further below).

In $\mathbb{R}^4$, the *length* of the vectors remain invariant during parallel displacement. In fact,

$$\boldsymbol{A} \cdot \boldsymbol{A} = A^i A^k g_{ik}$$

$$\mathrm{d}(\boldsymbol{A} \cdot \boldsymbol{A}) = \mathrm{d}g_{ik} A^i A^k + 2 g_{ik} A^i \mathrm{d}A^k \quad (g_{ik} = g_{ki})$$

$$\begin{aligned}(\mathrm{d}(\boldsymbol{A} \cdot \boldsymbol{A}))_\parallel &= (\mathrm{d}g_{ik})_\parallel\, A^i A^k - 2 g_{ik} A^i \Gamma^k_{lm} A^l \mathrm{d}y^m \quad ((\mathrm{d}g_{ik})_\parallel = \mathrm{d}g_{ik}) \\ &= (\mathrm{d}g_{il} - 2 g_{ik} \Gamma^k_{lm} \mathrm{d}y^m)\, A^i A^l \\ &= (g_{il,m} - g_{ik} \Gamma^k_{lm} - g_{lk} \Gamma^k_{im})\, \mathrm{d}y^m A^i A^l \\ &\quad \text{(exchange } i \text{ and } l \text{ in the 3rd term)} \\ &= g_{il;m} \mathrm{d}y^m A^i A^l = 0 .\end{aligned}$$

2.11. Symmetry and Antisymmetry Properties of Curvature and Torsion. Definition of the two Curvatures. Comparison between $\mathbb{M}^4$, $\mathbb{R}^4$ and $\mathbb{E}^4$.

Torsion is a vector

$$\begin{aligned}\Omega^i &= \frac{1}{2} (\Gamma^i_{jk} - \Gamma^i_{kj})\, (\mathrm{d}y^k \delta y^j - \mathrm{d}y^j \delta y^k) \\ &\equiv \tau^i_{jk}\, (-\mathrm{d}^2\sigma^{jk}).\end{aligned}$$

τ^i_{jk}, as well as $\mathrm{d}^2\sigma^{jk}$ are antisymmetric in j and k.

Because of the symmetry of the Γ's in $\mathbb{R}^4$, $\tau^i_{jk}=0$ in $\mathbb{R}^4$. The curvature depends on the Riemann-Christoffel tensor

$$R^l_{kpq}\equiv\Gamma^l_{kp,q}-\Gamma^l_{kq,p}+\Gamma^s_{kp}\Gamma^l_{sq}-\Gamma^s_{kq}\Gamma^l_{sp}$$

and yields the tensor called

curvature of rotation $\Omega^l_k\equiv R^l_{kpq}(-\mathrm{d}\sigma^{pq})$.

We see that R^l_{kpq} is antisymmetric in p and q.

We can sum up on $l=k$, which defines the ***homothetic curvature***

$$\Omega\equiv\Omega^k_k=R^k_{kpq}(-\mathrm{d}\sigma^{pq})\,.$$

In $\mathbb{R}^4$, $\Omega=0$. In $\mathbb{E}^4$, $\Omega=0$ and $\Omega^l_k=0$. This allows to build up the following table p. 33.

2.12. Identities Concerning Curvature in $\mathbb{R}^4$, and Connected Tensors

1⁰ If we put $i=k$,

$$R^i_{ipq}=\frac{\partial}{\partial y^q}\begin{Bmatrix}i\\ip\end{Bmatrix}-\frac{\partial}{\partial y^p}\begin{Bmatrix}i\\iq\end{Bmatrix}+\begin{Bmatrix}s\\ip\end{Bmatrix}\begin{Bmatrix}i\\sq\end{Bmatrix}-\begin{Bmatrix}s\\iq\end{Bmatrix}\begin{Bmatrix}i\\sp\end{Bmatrix}$$

(exchange i for s in the sums)

$$=\mathrm{d}^0-\mathrm{d}^0+\begin{Bmatrix}i\\sp\end{Bmatrix}\begin{Bmatrix}s\\iq\end{Bmatrix}-\begin{Bmatrix}i\\sp\end{Bmatrix}\begin{Bmatrix}s\\iq\end{Bmatrix}$$

$$=\frac{\partial}{y^q}\begin{Bmatrix}i\\ip\end{Bmatrix}-\frac{\partial}{\partial y^p}\begin{Bmatrix}i\\iq\end{Bmatrix}.$$

Now

$$\begin{Bmatrix}i\\ip\end{Bmatrix}=\frac{1}{2}g^{im}(g_{pm,i}+g_{im,p}-g_{ip,m})$$

$$=\frac{1}{2}g^{im}g_{pm,i}-\frac{1}{2}g^{mi}g_{ip,m}+\frac{1}{2}g^{im}g_{im,p}$$

$$=\frac{1}{2}g^{im}g_{pm,i}-\frac{1}{2}g^{im}g_{mp,i}+\frac{1}{2}g^{im}g_{im,p}$$

$$=\frac{1}{2}g^{im}g_{im,p}=\frac{1}{2}\frac{g_{,p}}{g}=\frac{1}{2}\frac{\partial}{\partial y^p}\quad(\ln g)$$

where $g\equiv\|g_{ik}\|$, $g^{ik}\equiv\dfrac{g_{ik}}{g}$.

	symmetries of g_{ik}	symmetries of Γ^i_{kl}	Absol. Differe- rent. Dg_{ik}	Ordinary Deri- vative $g_{ik,l}$	Cova- riant Deri- vative $g_{ik;l}$	Tor- sion	R^i_{kpq}: homothetic curvature	R^i_{kpq}: curvature of rotation	$\oint dm$	$\oint de_k$
$\mathbb{M}^4$	$g_{ik} \neq g_{ki}$	$\Gamma^i_{kl} \neq \Gamma^i_{lk}$	$Dg_{ik} \neq 0$	$\neq (e_i \cdot e_k)_{,l}$	$\neq 0$	$\neq 0$	$\Omega^i_k \neq 0$	$\Omega \neq 0$	$\neq 0$	$\neq 0$
							$R^i_{kpq} \neq 0$			
$\mathbb{R}^4$	$g_{ik} = g_{ki}$	$\Gamma^i_{kl} = \Gamma^i_{lk}$	$Dg_{ik} = 0$	$= (e_i \cdot e_k)_{,l}$	$= 0$	$= 0$	$\Omega^i_k \neq 0$	$\Omega = 0$	$= 0$	$= 0$
							$R^i_{kpq} \neq 0$			
$\mathbb{E}^4$	$g^i_k = \delta^i_k$ $g_{ik} = g_{ki}$	$\Gamma^i_{kl} = \Gamma^i_{lk} = 0$	$Dg_{ik} = 0$	$= (e_i \cdot e_k)_{,l}$	$= 0$	$= 0$	$\Omega^i_k = 0$	$\Omega = 0$	$= 0$	$= 0$
							$R^i_{kpq} = 0$			

Hence

$$R^i_{ipq} = \frac{1}{2}\left(\frac{\partial^2(\ln g)}{\partial y^q \partial y^p} - \frac{\partial^2(\ln g)}{\partial y^p \partial y^q}\right) = 0 .$$

There follows from this identity: $\Omega \equiv \Omega^i_i = -R^i_{ipq} d\sigma^{pq} = 0$.

2⁰ (a) If we make $i = q$, we obtain the

Ricci tensor $R_{kp} \equiv R^i_{kpi} = \ldots = R_{pk}$.

Moreover, with $R_{ikpq} \equiv g_{il} R^l_{kpq}$, we obtain a symmetry between the pairs ik and pq.

(b) If we define $R^k_p \equiv g^{ik} R_{ip}$ from R_{kp}, and if we put $k = p$, the *scalar curvature* $R \equiv R^k_k = g^{ik} R_{ik}$ is defined.

3⁰ (a) One defines

Einstein's tensor $S^i_k \equiv R^i_k - \frac{1}{2}\,\delta^i_k R$

and proves that

$$\nabla_i S^i_k \equiv 0 .$$

(b) The inverse question arises: which is the most general tensor C^i_k in $\mathbb{R}^4$, satisfying

$$\nabla_i C^i_k \equiv 0 ?$$

The answer was given by E. Cartan. It is what we call

Cartan's tensor $C^i_k = \alpha\left[R^i_k - \frac{1}{2}\,\delta^i_k\,(R - 2\lambda)\right]$

where α and λ are two constants.

4⁰ We leave it to the reader to prove

Bianchi's identities $\nabla_i R_{klpq} + \nabla_k R_{lipq} + \nabla_l R_{ikpq} = 0$.

2.13. Divergence, Curl and d'Alembert's Operator. Harmonic Coordinates

The *divergence* of a contravariant vector is its contracted covariant derivative $\operatorname{Div} \boldsymbol{A} = \nabla_i A^i$.

In $\mathbb{R}^4$, with $\Gamma = \{\}$, one has $\Gamma^k_{ak} = \ldots = \frac{1}{2}\, g^{ik} \frac{\partial g_{ik}}{\partial y^a}$ and, after a few calculations, we get

$$\operatorname{Div} \boldsymbol{A} = \frac{1}{\sqrt{-g}}\,(\sqrt{-g}\,A^i)_{,i} .$$

If $A^i = g^{ik}\varphi_{,k}$ where $\varphi_{,k}$ is called gradient of the scalar $\varphi(y)$, we can define d'Alembert's operator as applied on φ as the divergence of A^i,

$$\Box\varphi \equiv \frac{1}{\sqrt{-g}}(\sqrt{-g}\,g^{ik}\varphi_{,k})_{,i}.$$

It is an invariant.

We can put $\Box\varphi = g^{ik}\varphi_{ik}$. The tensor φ_{ik} is then nothing but the covariant derivative of the gradient of φ

$$\varphi_{ik} = \varphi_{,i,k} - \Gamma^l_{ik}\varphi_{,l} = \varphi_{ki}.$$

The identification of this new expression of $\Box\varphi$ with the above definition implies that

$$\Gamma^l \equiv g^{ik}\Gamma^l_{ik} = -\frac{1}{\sqrt{-g}}(\sqrt{-g}\,g^{lj})_{,j}$$

independant of φ. Thus, we could verify this relation directly.

Harmonic or isotherm coordinates ξ^i are made out of four independent solutions of

$$\Box\varphi = 0, \quad \text{i.e.} \quad \Box\xi^i = 0 \quad \text{with} \quad \xi^i_{,k} = \delta^i_k,$$

which are submitted in general to certain conditions at the limits (at ∞). We notice that for these coordinates the Γ^i's vanish and that we are not far away from (pseudo-)euclidicity.

The following operation on a vector $\boldsymbol{A}$ is called *curl* of $\boldsymbol{A}$:

$$\begin{aligned}(\mathrm{Rot}\,\boldsymbol{A})_{ik} &\equiv \nabla_i A_k - \nabla_k A_i\\ &= A_{k,i} - A_{i,k} = -(\mathrm{Rot}\,\boldsymbol{A})_{ki}.\end{aligned}$$

With a twice covariant tensor a_{ik}, two expressions can be formed:

$$\frac{1}{2}(a_{ik} \pm a_{ki}) \equiv \left\langle \begin{matrix} a_{\underline{ik}} \\ a_{\underset{\smile}{ik}} \end{matrix} \right. .$$

The first one $a_{\underline{ik}}$ is called the symmetric part, the second one $a_{\underset{\smile}{ik}}$ antisymmetric part. The curl equals twice the antisymmetric part of the covariant derivative of a covariant vector.

2.14. Geodesics in $\mathbb{R}^4$

It is the line defined by

$$\delta \int_A^B \mathrm{d}s = 0$$

passing through two fixed points A and B defined by the quadruples $y(A)$ and $y(B)$.

We can extract $2\mathrm{d}s\, \delta(\mathrm{d}s) = \ldots$ from $\mathrm{d}s^2 = g_{ik}\mathrm{d}y^i\mathrm{d}y^k$, and with ordinary means, we can find:

a first form of differential equation of the geodesic

$$\frac{\mathrm{d}^2y^i}{\mathrm{d}s^2} + \left\{ {i \atop pq} \right\} \frac{\mathrm{d}y^p}{\mathrm{d}s} \frac{\mathrm{d}y^q}{\mathrm{d}s} = 0$$

and if we define a "velocity"

$$u^k \equiv \frac{\mathrm{d}y^k}{\mathrm{d}s},$$

we find after a few calculations

a second form of the geodesic equation

$$u^k \nabla_k u^i = 0 .$$

2.15. Non-Riemannian Geometry

2.15.1. Non-Riemannian Manifolds

The use of a Riemannian manifold implies on the one hand that very symmetry and was on the other hand the starting point adopted originally by Einstein when he founded the (so-called generally relativistic) theory of gravitation (cf. Einstein's law of gravitation in sec. 3.2). A consequence of this use is that no skew-symmetric entities could be extracted from the theory itself, since a skew-symmetry can never be the consequence of a regular symmetry (whereas a regular symmetry can follows from a skew-symmetry by even numbers of operations). Now electromagnetism is typically skew-symmetric: the only regular symmetry encountered in electromagnetism is to be found in the distribution of energy (energy-momentum tensor, cf. 3.2.3). Therefore the original

Einsteinian (so-called generally relativistic, or rather generally covariant) theory of gravitation is uniquely a theory of the gravitational interaction, it cannot yield an explanation or description of the electromagnetic interaction, unless electromagnetism is artificially introduced and not, as one might wish, implied in the theory itself.

A theory which would yield implicitely both the gravitational and the electromagnetic interactions together is called a unitary theory. Attempts have been made, mainly by Einstein himself, to "unifiy" these interactions. Each attempt has led to a so-called unified-field theory. None of them, however, has been found to be the final and generally acceptable unitary theory. We shall therefore not try to describe the various attempts in detail but content ourselves with mentioning that Einstein himself tried mainly along the line suggested by above remark: to "generalize" general relativity to a description made by means of a variety V_4 in which the affine connection is not symmetric; we shall see further below how symmetric and skew-symmetric parts can be extracted from it (cf. sec. 2.11.1), from which attempts were made to interprete electromagnetism from the skew-symmetric one. (See however sec. 7.3 for further comments.)

2.15.2. Classification Obtained from the Affine Connection

Having shown that $\Gamma^i_{kl} = \left\{ {i \atop kl} \right\}$ in $\mathbb{R}^4$, i.e. the affine connection coincides with the Christoffel symbol of a Riemannian space, we shall henceforth use the notation A^i_{kl} for the coefficients of an affine connection not necessarily referring to a Riemannian space. The only property of a general differentiable manifold V_n is then its affinity. The latter is not a tensor. In V_n, its transformation rule can be proved to read as follows[1] (cf. sec. 2.3):

$$A'^i_{kl} = A^d_{ef} a'^i_d a^e_k a^f_l + \frac{\partial a^d_l}{\partial y'^k} a'^i_d \tag{1}$$

by a change of coordinates $y \to y'(y)$. Call $\underset{I}{A}{}^i_{kl}$, resp. $\underset{I}{A}{}'^i_{kl}$ the coefficients A^i_{kl} appearing in (1): Equ. (1) remain unchanged if they are replaced

[1] We shall not prove it here. See for instance A. Lichnerowicz, Théories relativistes de la gravitation et de l'électromagnétisme, Masson & Cie, Paris 1955, p. 238, equ. 57–5.

by

$$\underset{II}{A}{}^{i}_{kl} = \underset{I}{A}{}^{i}_{kl} + \Omega^{i}_{kl} \qquad (2)$$

(& same for A')

where Ω^{i}_{kl} is a tensor of 3d which of course transforms according to

$$\Omega'^{i}_{kl} = \Omega^{d}_{ef} a'^{i}_{d} a^{e}_{k} a^{f}_{l} . \qquad (2\text{a})$$

In V_n, a "covariant" derivative e.g. of a vector B^i shall be defined by

$$B^{i}_{;l} = B^{i}_{;l} + A^{i}_{kl} B^{k} . \qquad (3)$$

If a metric exists in V_n, its covariant derivative will be as above

$$g_{ik;l} = g_{ik,l} - A_{ikl} - A_{kil} \; (A_{ikl} \equiv g_{im} A^{m}_{kl}) . \qquad (4\text{a})$$

In case

$$g_{ik,l} - A_{ikl} - A_{kil} = 0 , \qquad (4\text{b})$$

i.e. if the metric is covariantly constant with regards to the affine connection, this connection itself and the transport it allows will be said to be metrical, whereas in case

$$g_{ik;l} \equiv \varphi_{ikl} \neq 0 , \qquad (4\text{c})$$

where, in virtue of the symmetry of the g_{ik}, the "inhomogeneity" φ_{ikl} is necessarily symmetric in its two first indices:

$$\varphi_{ikl} = \varphi_{kil} , \qquad (5)$$

the affine connection and the transport will then be nonmetrical. Equ. (4b) and its consequences constitute Ricci's lemma.

Now, there are n^3 coefficients of connections, but only $\frac{n^2}{2}(n+1)$ equations in Ricci's lemma (4b). So Ricci's lemma does not allow to determine all the coefficients of a metrical affine connection; $\frac{n^2}{2}(n-1)$ magnitudes remain undetermined and from (2) they constitute a tensor of rank 3.

If the A_{ikl} are assumed symmetric in their two last indices:

$$A_{ikl} = A_{ilk} \qquad (6)$$

they are found identical with the Christoffel symbols (from (4b))

$$A^i_{kl} = g^{im} \Gamma_{mkl} \,. \tag{7}$$

But then all the coefficients

$$\underset{II}{A}_{ikl} \equiv \Gamma_{ikl} + \beta_{ikl}$$

are solutions of Ricci's lemma (4b) as long as the 3d rank tensor β_{ikl} is skew-symmetric in its two first indices

$$\beta_{ikl} = -\beta_{kil} \,. \tag{8}$$

Condition (8) is therefore necessary and sufficient to fulfil Ricci's lemma, i.e. to define a metric connection. The β will be called generalized Ricci-coefficients. Thus: the metric connection belonging to a metric g_{ik} is the sum of the corresponding Christoffel symbols and the generalized Riccicoefficients.

Since the β are skew-symmetric in their first indices, they must be–as far as they do not vanish–non-symmetric in their last indices:

$$\beta_{ikl} \neq \beta_{ilk} \,.$$

They can be expressed as a linear combination of the torsion tensor τ_{ikl}:

$$\Omega_{ikl} = \beta_{ikl} = -\tau_{kil} - \tau_{lik} + \tau_{ikl} ;$$

where

$$\tau_{ikl} = -\tau_{ilk}$$

and

$$\tau_{ikl} = -\beta_{ilk} + \beta_{lki} + \beta_{ikl} \,.$$

Hence, every solution of Ricci's lemma which does not coincide with the Christoffel symbols describes a connection which does not fulfil the closure condition along an infinitesimal parallelogram

If, on the contrary, the Christoffel symbols are combined with an additive tensor ϱ_{ikl} which is symmetric in its first indices:

$$A_{ikl} = \Gamma_{ikl} + \varrho_{ikl}, \quad \varrho_{ikl} = \varrho_{kil} \,,$$

Ricci's lemma is not satisfied, but

$$g_{ik;l} = -\varrho_{ikl} - \varrho_{kil} = -2\varrho_{ikl} \equiv \varphi_{ikl} \,. \tag{10}$$

Then, in the most general case, equ. (2) can be explicited in the form

$$A_{ikl} = \Gamma_{ikl} + \Omega_{ikl} = \Gamma_{ikl} + \beta_{ikl} + \varrho_{ikl}\,, \tag{11}$$

where the first part Γ_{ikl} is the Christoffel symbol referring to the metric g_{ik}, the second part β_{ikl} is concomitant of the torsion and the third one, or rather $-2\varrho_{ikl}$, is the inhomogeneity in the covariant derivative of the metric.

The expression (11) allows a classification of the geometries of a V_n possessing a metric g_{ik}. It rests on the consideration of the displacement of vectors, for which a straight forward calculation based on (3) and (4a) will show, that–given a line $x^l = x^l(\tau)$ in terms of a parameter τ and given a vector B^i displaced along this line–the following expressions are equal: Parameter τ and given a vector B' displaced along this line, the following expressions are equal:

$$(B^i B^k g_{ik})_{;l} \frac{\mathrm{d}x^l}{\mathrm{d}\tau} = (B^i B_i)_{;l} \frac{\mathrm{d}x^l}{\mathrm{d}\tau} = (B^i_i)_{,l} \frac{\mathrm{d}x^l}{\mathrm{d}\tau} \equiv B^2/(x)\,.$$

There are three important cases:

(i) The simplest assumption yields the so-called Riemann-Christoffel or *infinitesimal parallel displacement*. It is the case where

$$\varrho_{ikl} = 0, \quad \beta_{ikl} = 0\,.$$

At the same time, the torsion vanishes, for

$$A_{ikl} - A_{ilk} = 0\,.$$

However, the displacement is not parallel at long distance, because the Riemann-Christoffel tensor of curvature

$$R^i_{kml} \equiv -\Gamma^i_{kl,m} + \Gamma^i_{km,l} - \Gamma^r_{kl}\Gamma^i_{rm} + \Gamma^r_{km}\Gamma^i_{rl}$$

does not vanish.

Nevertheless, the parallel displacement allows, according to Levi-Cività, a distant comparison between scalars and consequently between proper magnitudes in the physical sense of self-measurable magnitudes (e.g. length of vectors).

Indeed, if $x^l = x^l(\tau)$ defines a line in terms of a parameter τ, and a vector B^i be displaced along this line, the validity of Ricci's lemma allows to write

$$B^i B^k g_{ik;l} \frac{\mathrm{d}x^l}{\mathrm{d}\tau} = 0$$

for

$$B^i_{;l}\frac{dx^l}{d\tau}=0\,.$$

In this case (i), the Riemann-Christoffel tensor of curvature possesses all Ricci-symmetries

$$R_{ikmn}=-R_{iknm}=-R_{kimn}=R_{mnik};$$

$$R_{iklm}+R_{ilmk}+R_{imkl}=0$$

and does not in general vanish.

(ii) The *displacement* is *simply metrical* if in (11)

$$\varrho_{ikl}=0$$

but $\beta_{ikl}\neq 0$, meaning that the torsion does not vanish

$$A_{ikl}-A_{ilk}=2\tau_{ikl}\neq 0\,.$$

The tensor of curvature will in this case be called G^i_{klm} and is found equal to

$$\begin{aligned}G^i_{klm}&\equiv -A^i_{kl,m}+A^i_{km,l}-A^r_{kl}A^i_{rm}+A^r_{km}A^i_{rs}\\&=R^i_{kml}-\Omega^i_{kl;m}+\Omega^i_{km;l}-\Omega^r_{kl}\Omega^i_{rm}+\Omega^r_{km}\Omega^i_{rl}\,.\end{aligned}$$

It does not in general vanish and possesses only the following symmetries:

$$G_{iklm}=-G_{ikml}=-G_{kilm}\,.$$

A comparison of displaced proper magnitudes is still possible:

$$B^2/(x)=0$$

for, again,

$$B^i_{;l}\frac{dx^l}{d\tau}=0\,.$$

(iii) In one case of this simply metrical displacement the curvature tensor does vanish, which allows for comparisons between directions and not only between proper magnitudes. The displacement is then called Einsteinian parallel displacement at distance. For this to be seen, let us introduce an orthonormal field of "n-pods" h^i_D (contravariant components), resp. h_{Dk} (covariant components ($D=1, 2, \ldots n$) and write the metric

$$g_{ik}=h_i^D h_{Dk}\,.$$

For that particular field, Einstein's lemma

$$h^D_{i;l}\equiv h^D_{i,l}-h^D_k \Delta^k_{il}=0$$

applies. Ricci's lemma is of course also valid. Then, from Einstein's lemma, the Einstein coefficients (then called Δ^i_{kl}) of the Einstein displacement are found equal to

$$\Delta^i_{kl}=A^i_{kl}=h^{Di}h_{Dk,l}=-h^{Di}_{,l}h_{Dk}\,,$$

or, to have it in the form (11),

$$\Delta_{ikl}=\Gamma_{ikl}+h_i^D h_{Dk;l}\,,$$

showing what β_{ikl} (then called Ricci's rotation coefficients γ_{ikl}) consists in

$$\gamma_{ikl}=h_i^D h_{Dk;l}=-h^D_{i;l}h_{Dk}\,,$$

and yielding for the 3d rank torsion tensor

$$\tau_{ikl}=\frac{1}{2}\,h_i^D\,(h_{Dk,l}-h_{Dl,k})\,.$$

Einstein's lemma warrants the integrability of the displacements. The integrability condition being summarized in the vanishing of the curvature:

$$G^i_{klm}=-\Delta^i_{kl,m}+\Delta^i_{km,l}-\Delta^r_{kl}\Delta^i_{rm}+\Delta^r_{km}\Delta^i_{rl}\,.$$

(Einstein)

(iii) The most general case assumes (beside $\beta_{ikl}\neq 0$)

$$\varrho_{ikl}\neq 0\Leftrightarrow\Omega_{ikl}+\Omega_{kil}\neq 0\,.$$

Comparison is redered impossible not only between directions, but also between proper magnitudes, for then the square of B^i is found to change along the line $x^l=x^l(\tau)$ according to

$$B^2/(x)=B^iB^kg_{ik;l}=-2\varrho_{ikl}B^iB^k\frac{dx^l}{d\tau}\,.$$

From this, there follows that the absolute values themselves of vectorial properties of systems are found different if they are analyzed along different lines of approach from an initial position. For physical systems embeded in space-time, this means that the proper magnitudes attached to them depend upon the past history of these systems: If at a time $t=0$ two (pointlike) systems situated at the same point P_0 of the subspace V_3 had all the same proper magnitudes but if they from then on are transported, according to Levi-Cività's definition of a parallel displacement, along two different lines into a unique point P_a where they meet at the same time $t=a$, the proper magnitudes of the two systems will differ. These magnitudes possess biogrammatic features. This can be interpreted by saying, that the part of Ω_{ikl} which is symmetric in the two first indices, viz. ϱ_{ikl}, incorporates the temporal (historical) behaviour of the physical processes. This was discovered by Einstein as early as 1918 in a discussion with Hermann Weyl about conformal geometry (cf. 7.3.3).

The only symmetry remains with the Riemannian tensor G in this general case (iii) is

$$G_{iklm} = -G_{ikml} \,.$$

The skew-symmetry of the two first indices is no more valid; it warranted the integrability for the case of the validity of Ricci's lemma.

2.15.3. Further Classification Obtained from the Relation between Affine Connection and Curvature Tensor

This allows a more detailed classification than under 2.11.1.

2.15.3.1. Most general case:

$$g_{ik;l} = -2\varrho_{ikl} = \varphi_{ikl} \neq 0$$

$$A^i_{kl} = A^i_{lk} \neq 0 \,.$$

The curvature tensor admits only of the trivial symmetry

$$R^i_{klm} = -R^i_{kml}$$

and neither the curvature of rotation nor the torsion vanish:

$$\Omega_{lk} \neq 0, \quad \tau^i_{kl} \neq 0 \neq \Omega^i \,.$$

2.15.3.2. The displacement is metrical; then

$$g_{ik;l} = 0;$$

however the *torsion does not vanish*:

$$\tau^i_{kl} \neq 0 \neq \Omega^i .$$

Then the curvature of rotation vanishes

$$\Omega_{lk} = 0 .$$

2.15.3.3. The displacement is non-metrical but the affine connection is symmetric:

$$g_{ik;l} = -\varphi_{ikl} \neq 0 .$$

However,

$$\tau^i_{kl} = 0 = \Omega^i .$$

Furthermore

$$A^i_{kl} - A^i_{lk} = \frac{1}{2} g^{ir}[kl, r] + g^{ir} \varrho_{rkl}$$

and

$$R^i_{klm} + R^i_{lmk} + R^i_{mkl} = 0 .$$

2.15.3.4. Riemannian spaces are at hand when

$$g_{ik;l} = 0, \quad \tau^i_{kl} = 0 = \Omega^i$$

$$A^i_{kl} = \begin{Bmatrix} i \\ kl \end{Bmatrix} = A^i_{lk}, \quad \varrho_{ikl} = 0 = \varphi_{ikl}$$

$$R^i_{klm} + R^i_{lmk} + R^i_{mkl} = 0$$

$$R_{ikmn} = -R_{kimn} = R_{kinm} = R_{mnik} .$$

2.15.3.5. Spaces admitting of a parallel displacement at distance require that

$$R^i_{klm} = 0 ;$$

however they admit of a torsion

$$\Omega^i \neq 0 ;$$

and, in these spaces,

$$A^i_{kl} = h^i_D h^D_{kl}, \quad \text{hence} \quad g_{ik;l} = 0$$

2.15.3.6. Flat spaces are defined as having neither curvature of any kind, nor torsion:

$$R^i_{klm} = 0, \quad \Omega^i = 0 .$$

They are often assumed as the asymptotic behaviour of spaces where a gravitational field is produced by the presence of matter.

3. FUNDAMENTALS OF GRG

3.1. Geometrization of Matter

Newton had elaborated a dynamics in which the elements of matter act upon one another by means of forces which create in the course of motion modifications considered different from uniformity.

Descartes before him had had the idea of filling up space with vortices creating a kinematical state whose evolution would be determined for ever and which would be *the* structure of the world. He did not assume any forces and his theory though scarcely elaborated from the mathematical point of view tended to "geometrize" the problem of matter rather than to "dynamize" it.

Einstein went as far as he thought possible on this second path. One has since gone another important step further in the same direction (Wheeler's theory of geons and of "worm holes"). Thus there are two steps:

First, a distribution of matter creates the geometrical variety of the world.

Second, the idea that there is matter independently of space is an illusion. There is but a well defined variety, the local properties of which can be "naïvely" interpreted as "present matter".

Along the first step, the problem of Matter and Geometry is put in a rather awkward manner, for the distribution is a clear notion only by means of a distribution function which possesses the nature of a density and which therefore will have to be a function of the points y of the variety, that is, finally of the coordinates y^k; but the variety as a structure of points is itself determined by means of coordinates appearing in the metric $g_{ik}(y)$, i.e. by functions of points; hence, one does not know clearly where to begin: by choosing the coordinates? By choosing the g_{ik}? By choosing the density? . . .

Apart from this, once the density is known, it will be a particular one which cannot be exchanged for another one as considered as a function of the four coordinates of space-time; hence it is submitted to an absolute determinism, meaning that it is completely determined everywhere and for all times. The question immediately arises why this particular density rather than another one should be realized in the universe? There are only two possible answers to that question. One consists in assuming a Creator transcending physical Nature and made responsible for the choice. According to the second one, the realized function presents no "peculiar peculiarity" but is the sole one for which there is no reason whatsoever why another function should be preferable, in other words, it is the best of all possible ones.[1] Yet one glance around us is enough to allow the claim that the distribution has peculiarities, since there are stars of various dimensions at variable distances and all sorts of distinct physical objects. Our own existence is in itself a peculiarity. The one way not to be compelled to consider each neighbourhood (different from all others) as a particularity peculiar to a singular distribution is to interpret it as a pure fluctuation in a regular distribution. If so, then either everything we observe is nothing but the effect of fluctuations around a state in itself unobservable, or it is the aspect of a universe created as it is rather than otherwise, but physics must refrain from searching reasons for this creation since they will have to be beyond the purely physical behavior of things. Under these conditions, the universe appears as it does only as a result of a play the author of which does not concern physics.

By the way, if we believe that we know the distribution of matter, this distribution is from the point of view of mathematics a mere function of the coordinates which are four series of numbers. The projection of this functional relation onto physical objects requires a convention establishing a correspondence between each quadruple y^i and a physical situation, i.e. a physically noticeable event. This implies quite a problem concerning

[1] This is a Leibnizian position which, anyhow, again leads to the belief in a perfect God.

physical interpretation, namely as to how measurements of distributed properties are to be made by means of parameters, themselves readable on graduations which are assumed to specify the time and the place of the event.

The distributed properties, in their turn build entities which are more complex than a mere density in the ordinary sense of elementary physics. They appear already in special relativity where mass and energy are inseparable. Thus, if one talks about density as a mass distribution, the distribution of all forms of energy must be taken along with it and vice versa. Moreover, space-time is a vaster continuum than three-dimensional space of our usual vision and current densities will have to be considered too. Special relativity suggests how to combine these densities, for it teaches that in electrodynamics energy parts into: a density w, Poynting's flux $\vec{S}$ and Maxwell's tensions (tensor $\underline{\underline{M}}$) to be set together as follows in an *energy-momentum tensor*.

$$\underset{\circ}{\tau}^{lk}=\left(\begin{array}{c|c} w & i\vec{s} \\ \hline i\vec{s} & \underline{\underline{M}} \end{array}\right) \quad (i\equiv\sqrt{-1})$$

where the value 0 of l or k is attributed to the temporal coordinate of the chosen Galilean system, whereas the index o under $\underset{\circ}{\tau}$ refers to the Galilean system of coordinates. From this, Einstein conceived the possibility of considering an $\mathbb{R}^4$ and the (pseudo-)Euclidean space $\mathbb{E}^4(\boldsymbol{P})$ tangent in each point $\boldsymbol{P}$ of $\mathbb{R}^4$ where a $\underset{\circ}{T}^{ik}(\boldsymbol{P})$ is to be found. In a point $\boldsymbol{Q}$ of $\mathbb{R}^4$ infinitely near to $\boldsymbol{P}$, which is properly speaking not a point of $\mathbb{E}^4(\boldsymbol{P})$, one can take as the value of T^{ik} the value $\underset{\circ}{T}^{ik}(\boldsymbol{P}+\mathrm{d}\boldsymbol{P})$, where $\boldsymbol{P}+\mathrm{d}\boldsymbol{P}$ is the projection of $\boldsymbol{Q}$ in the tangent space $\mathbb{E}^4(\boldsymbol{P})$. Moreover, T^{ik} shall contain, not only the electromagnetic energy-momentum τ^{ik} but every kind of energy-momentum. In particular, if there is nothing in space-time but massive matter in a state of powder without any electric charge, like a gas or a cloud of dust, the particles of which possess quadrivelocities u^k (again defined from tangent space $\mathbb{E}^4(\boldsymbol{P})$), one has to deal with a *material energy-momentum tensor*.

$$\underset{\circ}{M}^{ik}=\mu c^2 u^i u^k\,,$$

where $\underset{\circ}{\mu}$ is the density of mass in the usual sense as measured at the point $\boldsymbol{P}$ in $\mathbb{E}^4(\boldsymbol{P})^0$.

If there are stresses P_{ik}, they have to be added; but it is clear that P^{0k} vanishes in the Galilean system of coordinates, for these internal stresses known from the physics of deformable bodies arise only from elasticity and viscosity.

Einstein understood that there must exist a relation between the metric and the distribution. Now the latter is characterized by a symmetric tensor of second rank about which it is known that in a Galilean system it satisfies equations of conservation

$$\underset{\circ}{\nabla}_k \underset{\circ}{T}_i^k = 0 \,.$$

So Einstein postulated that according to general covariance

$$\nabla_k T_i^k = 0 \,.$$

These equations will be called "conservation equations" in quotation marks, for we shall see later that there is no authentic conservation.

In other words, the problem is to find conditions which guarantee the vanishing of the covariant derivative of the energy-momentum tensor T_i^k as related to $\mathbb{R}^4$.

Besides, in the absence of any matter, space-time must be everywhere and always Euclidean[1] (g_{ik} reduces to $\pm\delta_{ik}$). There must be some relation between g_{ik} and T_{ik}, but in no case a finite functional one, for if it were, it would not allow for limit conditions (beside normalization). Hence, this relation will imply the derivatives of g_{ik}.

We must therefore look for a tensor of second rank built on the g_{ik} and their derivatives up to a suitable rank, such that when functionally related to T_i^k, it will automatically guarantee the "conservations". Moreover this relation must allow for the normalization, i.e. be such that, when T_i^k vanishes everywhere identically, the g_{ik}, solutions of differential equations, reduce to $\underset{\circ}{g}_{ik}$ with the signature $(+1, -1, -1, -1)$.

Cartan's tensor C_i^k is known to be the most general one obtained from the metric and its first and second derivatives and satisfying these "conservation equations". It is, by the way symmetric when written C_{ik} or C^{ik}.

[1] This condition of reduction to vacuum, sometimes called the normalization of the metric, seems at first sight evident. In fact, it seemed natural in the days when these considerations were developed for the first time. In our days, one is rather inclined to think that Minkowski-space itself is not necessarily the correct frame of description for the case where matter is absent altogether, and there are proposals to normalize the metric by means of a space obtained by fixing the metric as the one following from that solution of the field equations which describe space-time as void.

It seems therefore well founded to postulate
$C_i^k = \chi T_i^k$ within matter and energy.

In this formula χ is a universal constant.

It would then be enough to prove that $g_{ik} = \overset{\circ}{g}_{ik}$ follows from $C_i^k \equiv 0$.

Under these conditions, these would be the *field laws*.

In the exterior of matter-energy, $C_i^k = 0$ automatically, but this does not imply that the g_{ik} are constant. C_i^k takes the form

$$C_i^k = \alpha \left[R_i^k - \frac{1}{2}\, \delta_i^k (R - 2\lambda) \right],$$

where α and λ are constants. α can be incorporated into χ, so λ remains alone. Hence, there are two universal constants χ and λ.

At the origin, Einstein considered only the tensor S_i^k, which corresponds to the choice $\alpha = 1$, $\lambda = 0$. That is what is known under

Einstein's law of gravitation $\quad S_i^k = \chi T_i^k$

i.e. in detail

$$R_i^k - \frac{1}{2}\, \delta_i^k R = \chi T_i^k .$$

Beside this law (which is analogous to GMm/r^2 of Newtonian mechanics), Einstein believed that it was necessary to postulate a fundamental or "dynamical" equation, namely e.g. the equation of the geodesiclines in its differential form (analogous to "force = mass × acceleration" in Newtonian mechanics).

Now, in a theory of *pure gravitation* (i.e. in the absence of any other interaction), this has been found unnecessary. For, such a dynamical equation was in Newtonian mechanics used to determine the trajectory. But the trajectory is determined in GRG by its identity with a world-line, and world-lines are included in the quadridistribution of energy-momentum, since all mass is energy.

It would even be dangerous to postulate such a dynamical equation because when going from the case of pure gravitation over to the case where the equation is assumed valid also in the presence of several interactions, in particular of gravitation and electromagnetism, one risks either to find the wrong trajectory by keeping the same fundamental equation or to be compelled to exchange it for another fundamental equation of which the old one is nothing more than a particular case. (The latter is more or less what has happened in the artificial combination realized by "glueing" electromagnetism in its Maxwellian description on to gravitation).

One should much rather look for a way to extract "dynamics" from "geometry". This is done as follows:

Let us first consider pure gravitation and restrict ourselves to dust for which the material tensor M_i^k reads

$$M_i^k = \mathring{\mu} c^2 u_i u^k .$$

Its divergence must vanish:

$$\nabla_k M_i^k = 0 = \nabla_k (u_i u^k) = u_i \nabla_k u^k + u^k \nabla_k u_i .$$

Now dust is made out of incompressible particles. The substance of each particle is conserved. Hence, the divergence of the vector u^i vanishes (continuity equation): $\nabla_k u^k = 0$. So

$$u^k \nabla_k u^i = 0$$

which is the equation of the *geodesic.*

The proof has been given here for dust without inner stress only and is true only in the absence of electrically charged matter as well as of electromagnetic fields. Otherwise, if one assumes that it is enough to artificially glue electromagnetism onto gravitation, τ_i^k must be added to M_i^k. But for an electromagnetic field F_{ik} in the absence of electrically or magnetically polarizable substances,

$$\tau_i^k = \mathring{\varepsilon} \left(-F_{il} F^{kl} + \frac{1}{4} \delta_i^k F_{ml} F^{ml} \right)$$

where $\mathring{\varepsilon}$ is the constant of electric induction (this formula being the mere transcription into the generally covariant formalism of what was in special relativity written by means of τ_i^k and $\underset{\circ}{F}{}_i^k$). In virtue of Maxwell's equations, the covariant derivative $\nabla_k \tau_i^k$ reduces to:

$$F_{ik} \mathring{\varepsilon} \nabla_l F^{kl} = F_{ik} j^k$$

where j^k is the quadri-density of electric current. This is nothing but the Lorentz force density, whence

$$u^k \nabla_k u^i = -\mathring{\varepsilon} F_{ik} j^k .$$

Therefore charged particles *do not follow* the geodesics in the presence of an electromagnetic field, and Einstein's postulate about trajectories as geodesic lines was not as general as that!

Photons are the oldest known neutral particles. As a matter of fact, until the discovery of the neutron and of neutrinos, they were believed to be the only rigorously neutral particles. Atoms, molecules, planets are "particles" only at the levels of appropriate aproximations.

For photons and for neutrinos too,

$$ds = 0$$

for they move at velocity c and they have no mass at rest. But then no vector tangent to $y^i = y^i(s)$ can be defined because s loses its meaning as parameter. However, one can think of another parameter σ and, in order that $ds^2 = 0$, one can write

$$g_{ij} \frac{dy^i}{d\sigma} \frac{dy^j}{d\sigma} = 0 .$$

Then the equation of the geodesic still holds under a slightly modified form:

$$\frac{d^2 y^i}{d\sigma^2} + \Gamma^i_{jk} \frac{dy^j}{d\sigma} \frac{dy^k}{d\sigma} = 0 .$$

Summing up, the geometrization of matter consists in establishing a functional relation between the metric and the distribution, and this only law of the field–which in fact splits into several equations for tensor components–determines in principle everything.

The questions then arising are of two kinds: how does one practically write this functional relation called field equation for specific cases, and what are the general consequences of this functional relation independent from such specific cases. Answering such questions constitutes what might be called the geometro-dynamics of gravitation, or gravito-geometro-dynamics as distinguishable from a gravito-dynamics of Newtonian inspiration.

3.2. Einsteinian Gravito-dynamics or Gravito-geometro-dynamics

If we persist in believing that matter is something independent from the geometrical structure of space-time but capable of imposing that structure, it means that we do not follow the Cartesian ideal of a geometric model of the world up to its very end. Then the field equations are the expression of a "remaining dynamics", since geometry is determined by matter in a way similar to how motion was determined by forces acting on matter.

Let us notice, by the way, that we are confronted with a comparable situation in quantum field theory, i.e. the modern theory of particles. For, one can first assume that there are particles endowed with various properties. Then, one can look at particles as the quanta of a quantizable field for which field equations are postulated. Finally, one can say that "particle" is just a name to designate that which in a naive representation is believed to be responsible for what appeared to be collisions, dispersion phenomena and the like; but, there really are no isolated particles because one never sees anything else than microscopic catastrophies which are called collisions or what have you. Or still, we could say that the word "particle" is just a name used to designate a so-called element of a representation by groups of symmetry in the modern $SU(n)$ theory of particles, for instance that there is really nothing but the groups themselves which are mathematically conceived and axiomatically clarified.

Going back to gravito-dynamics, let us recall that the determination of motion by forces is, so to say, a fairly inelegant though convenient procedure for explicit calculations instead of saying that trajectories are the consequence of a variation principle for the action. One can then ask whether Einstein's equations of the gravitational field are also the (inelegant) expression of the consequences of a variation principle. This will be the case if it proves possible to give the exact form of an action functional from which they would follow. This is indeed possible. Action functionals are integrals calculated in domains where the integrand is multiplied by an integration element which in the course of changes of coordinates is to transform as required by Jacobi's theorem based on the use of the Jacobian.

The action itself should not depend on the coordinates if it is necessary and sufficient for the determination of this trajectory or for the establishment of the equations following from it, for the trajectories are world-lines which remain intrinsically unchanged whatever be the choice of coordinates.

As a matter of fact, all the equations which can be deduced from the variation principle shall yield more than just the trajectories. They will part into those which describe the motion of test-particles, into "conservation" equations and into equations of the field itself, according to whether the action shall be varied by means of arbitrary variations of the coordinates of a test particle, of the metric or of the field parameters.

However, to ensure that the action be an invariant towards changes of coordinates, the integrand itself must have the nature of a density. This is why we shall presently elaborate the formalism of what are called tensorial densities and study the construction of invariants.

3.2.1. Tensor-densities and Invariants

According to Jacobi's theorem, the element $\mathrm{d}V$ (e.g. $\mathrm{d}\overset{\circ}{V}=\mathrm{d}\overset{\circ}{y}^1\wedge\mathrm{d}\overset{\circ}{y}^2\wedge\wedge\ldots\wedge\mathrm{d}\overset{\circ}{y}^n$) of volume used in integrals transforms along coordinate changes $y'^l=y'^l(y)$ according to

$$\mathrm{d}V'=\left\|\frac{\partial y'^l}{\partial y^m}\right\|\mathrm{d}V\,.$$

Hence, it is not an invariant.
By defining, as further above,

$$a^{l'}{}_m\equiv\frac{\partial y'^l}{\partial y^m}\,,$$

and

$$a^l{}_{m'}\equiv\frac{\partial y^l}{\partial y'^m}$$

for the inverse transformation, $a^l_{m'}$, and $a^{l'}_m$ fix covariance and contravariance. Let us put

$$a\equiv\left\|\frac{\partial y^l}{\partial y'^m}\right\|\quad\text{and}\quad a'\equiv\left\|\frac{\partial y'^l}{\partial y^m}\right\|,\quad aa'=1$$

for the determinants. For example, $\mathrm{d}V'=a'\mathrm{d}V$.

Beside these determinants, another one is important:

$$g\equiv\|g_{ik}\|\quad\text{with}\quad\|\overset{\circ}{g}_{ik}\|=\begin{vmatrix}1&0&0&0\\0&-1&0&0\\0&0&-1&0\\0&0&0&-1\end{vmatrix}=-1\,.$$

This determinant of the metric transforms as follows:

$$g'=a^2g,\quad\text{or}\quad\frac{1}{a}\,g'=ag,\quad\text{i.e.}\quad a'g'=ag,$$

and

$$\sqrt{-g'} = a\sqrt{-g}, \quad (\sqrt{-g} > 0).$$

Let $A^{\cdots}_{\cdots}$ be a tensor with covariant and contravariant indices in arbitrary numbers. The magnitude

$$\mathfrak{A}^{\cdots}_{\cdots} = \sqrt{-g}\,A^{\cdots}_{\cdots}$$

is called the corresponding ***tensorial density.***

The simplest density is a scalar density

$$\mathfrak{A} = \sqrt{-g}A,$$

A being a scalar (an invariant). Since $A' = A$, $\mathfrak{A}$ transforms like $\sqrt{-g}$, i.e.

$$\mathfrak{A}' = a\mathfrak{A}.$$

But $\sqrt{-g}\,\mathrm{d}V \to a\sqrt{-g}\,a'\mathrm{d}V = \sqrt{-g}\,\mathrm{d}V$. Hence, there exists an invariant which can be used instead of the element of volume, viz.

$$\sqrt{-g}\,\mathrm{d}V.$$

In other words: the "volume density" is invariant. This explains the use of the word "density".

Now, in $\mathbb{R}^4$,

$$\nabla_l g_{ik} = 0;$$

hence,

$$\begin{aligned}\nabla_l \mathfrak{A}^{\cdots}_{\cdots} &= \nabla_l(\sqrt{-g}\,A^{\cdots}_{\cdots})\\ &= (\sqrt{-g}\,A^{\cdots}_{\cdots})_{,l}\\ &\quad + [\{\}\,\mathrm{A}^{\cdots}_{\cdots} \ldots]\sqrt{-g}\\ &= \mathfrak{A}^{\cdots}_{\cdots,l} - \frac{(\sqrt{-g})_{,l}}{\sqrt{-g}}\,\mathfrak{A}^{\cdots}_{\cdots} + \{\,\}\,\mathfrak{A}^{\cdots}_{\cdots}\\ &\quad + \ldots - \{\,\}\,\mathfrak{A}^{\cdots}_{\cdots} - \ldots.\end{aligned}$$

Furthermore,

$$\begin{aligned}\mathrm{d}g &= \sum(\mathrm{d}g_{ik})\,(\min g_{ik}) = g g^{ik}\mathrm{d}g_{ik}\\ &= -g g_{ik}\mathrm{d}g^{ik}\end{aligned}$$

and

$$\begin{Bmatrix} i \\ ki \end{Bmatrix} = \frac{1}{2} g^{im} (g_{mk,i} + g_{mi,k} - g_{ki,m})$$
$$= \frac{1}{2} g^{im} g_{mk,i} - \frac{1}{2} g^{im} g_{km,i} + \frac{1}{2} g^{im} g_{mi,k}$$
$$= \frac{1}{2} g^{im} g_{im,k}$$

whence

$$\frac{1}{2g} g_{,k} \equiv \frac{1}{2g} \frac{\partial g}{\partial y^k} = \begin{Bmatrix} i \\ ki \end{Bmatrix} = \begin{Bmatrix} i \\ ik \end{Bmatrix},$$

i.e.

$$\begin{Bmatrix} i \\ ik \end{Bmatrix} = (\ln \sqrt{-g})_{,k} .$$

Therefore, the true divergence, which is the contracted covariant derivative of a contravariant vector, is equal to

$$\operatorname{div} \boldsymbol{A} \equiv A^i_{;i} = A^i_{,i} + \begin{Bmatrix} i \\ ik \end{Bmatrix} A^k$$
$$= A^i_{,i} + A^l (\ln \sqrt{-g})_{,l}$$
$$= \frac{1}{\sqrt{-g}} (\sqrt{-g}\, A^i)_{,i}$$
$$= \frac{1}{\sqrt{-g}} \mathfrak{A}^i_{,i} .$$

In an analogous way, considering a twice contravariant antisymmetric tensor

$$A^{\underset{\smile}{ik}} = -A^{\underset{\smile}{ki}}$$

(the symbol $\smile$ is characteristic of the antisymmetry), we can define and compute the big divergence as the following contracted derivative

$$(\operatorname{Div} \underline{A})^i \equiv A^{\underset{\smile}{ik}}_{;k} = A^{\underset{\smile}{ik}}_{,k} + \Gamma^i_{mk} A^{\underset{\smile}{mk}} + \Gamma^k_{km} A^{\underset{\smile}{im}}$$
$$= A^{\underset{\smile}{ik}}_{;k} + \Gamma^k_{mk} A^{\underset{\smile}{im}} \quad (\text{for } \Gamma^i_{mk} A^{mk} = -\Gamma^i_{km} A^{km} = 0)$$
$$= A^{\underset{\smile}{ik}}_{,k} + \frac{(\sqrt{-g})_{,m}}{\sqrt{-g}} A^{\underset{\smile}{im}}$$

$$=\frac{1}{\sqrt{-g}}(\sqrt{-g}\,A^{\underset{\smile}{im}})_{,m}$$

$$=\frac{1}{\sqrt{-g}}\,\mathfrak{A}^{\underset{\smile}{ik}}{}_{,k}\,.$$

For a symmetric tensor $S^{\underline{kl}}=S^{\underline{lk}}$ there is no such simple formula for its big divergence, one must leave it at

$$S^k_{i;k}=\frac{1}{\sqrt{-g}}(\sqrt{-g}\,S^k_i)_{,k}-\frac{1}{2}\,g_{kl,i}S^{\underline{kl}}$$

where $S^k_i\equiv g_{il}S^{\underline{kl}}$.

Coming back to the (little) divergence div $\boldsymbol{A}$ of a vector, we can apply Gauss' theorem:

$$\oint A^i\sqrt{-g}\,\mathrm{d}S_i=\int A^i_{;i}\sqrt{-g}\,\mathrm{d}V,$$

where $\mathrm{d}S_i$, is the (covariant) vector dual to the element $\mathrm{d}S^{klm}$ of hypersurface at the limit of the domain V whose element is $\mathrm{d}V$:

$$\mathrm{d}S_i=\frac{1}{6}\,e_{iklm}\mathrm{d}s^{klm}$$

where e_{iklm} is the (pseudo-scalar) antisymmetric unit tensor of rank 4.

These formulae can be established first in a tangent Cartesian system of orthogonal coordinates and within an infinitesimal domain and generalized then to any coordinate system.

We shall apply them in the next section.

3.2.2. Origin of the Energy-momentum Tensor

Ordinary mechanics, for which we may use orthogonal cartesian coordinates q for which all indices will be put down, can be summarized in a variational principle, starting with an action

$$S=\frac{1}{ic}\int\mathfrak{L}(q,\dot{q})\,\mathrm{d}^4V$$

where $\mathrm{d}^4V=\mathrm{d}^3V\,ic\mathrm{d}t$, so that $L=\int\mathfrak{L}(\mathrm{d}V)^3$ is the ordinary Lagrangian and $\mathfrak{L}$ its density.

S is a functional built upon a space itself dedoubled in a q-space of configuration and a $\dot{q}$-space of velocities; the latter is not totally alien to configuration space since the $\dot{q}$'s are velocities and since a trajectory in configuration space determines the hodograph in a $\dot{q}$-space. The notion of a velocity can be generalized by considering $\dot{q}$ as the symbol of four derivatives $q_{,i}$ to be formed from any q. Then,

$$\delta S=0$$

yields the equations

$$\frac{\partial}{\partial q_i}\frac{\partial \mathfrak{L}}{\partial(q_{,i})}-\frac{\partial \mathfrak{L}}{\partial q}=0 \quad (i=1,\ldots 4)$$

which can be written

$$(\mathfrak{L}_{,q_{,i}})\;\mathfrak{L}_{,q}=0\,.$$

Then

$$\begin{aligned}\mathfrak{L}_{,i}&=\mathfrak{L}_{,q}q_{,i}+\mathfrak{L}_{,q_{,k}}q_{,k,i}\\&=(\mathfrak{L}_{,q_{,k}})_{,k}+\mathfrak{L}_{,q_{,k}}q_{,i,k}\\&=(\mathfrak{L}_{,q_{,k}}q_{,i})_{,k}\end{aligned}$$

or

$$\begin{aligned}\delta_{ik}\mathfrak{L}_{,k}-(q_{,i}\mathfrak{L}_{,q_{,k}})_{,k}&=0\\&=(\delta_{ik}\mathfrak{L}-q_{,i}\mathfrak{L}_{,q_{,k}})_{,k}\,.\end{aligned}$$

This relation takes on the form of an equation of *conservation*

$$(T_{ik})_{,k}=0$$

if we put

$$T_{ik}\equiv\mathfrak{L}\delta_{ik}-q_{,i}\mathfrak{L}_{,q_{,k}}\,.$$

Of course, if q is the symbol of a set of parameters q_α, one has to put

$$T_{ik}=\mathfrak{L}\delta_{ik}-q_{\alpha,i}\mathfrak{L}_{,q_{\alpha,k}}$$

where a summation over α is tacitly understood.

The above conservation relations can still be written

$$P_i\equiv\int T_{ik}\mathrm{d}S_k=\text{constant}\;|_i$$

where the word constant means invariable as a function of the parameter upon which the q_α may depend.

In general, one does not use within the frame of classical analytical dynamics a four-dimensional space-time ($i, k, \ldots, \alpha, \ldots = 1, 2, 3, 4$) but one separates time t and uses it as the independent parameter. But here, t (or ict) is one of the coordinates and all four coordinates are equivalent. So, one has to go over to a *homogeneous formalism* where the independent parameter is no more time itself, and to prove that, possibly under accessory conditions, this homogeneous formalism does not describe anything more than what is described by ordinary mechanics in the inhomogeneous formalism. We shall not do it here[1].

Going back to T_{ik}, let us assume the existence of an antisymmetric field $\Psi_{\underset{\smile}{ikl}}$ which vanishes at the limit of the domain of its definition. From the identity

$$\Psi_{\underset{\smile}{ikl},k,l}=0$$

and by applying Stokes' theorem, there follows that

$$\int \Psi_{\underset{\smile}{ikl},l}\mathrm{d}S_k=\frac{1}{2}\int (\Psi_{\underset{\smile}{ikl},l}\mathrm{d}S_k-\Psi_{\underset{\smile}{ikl},k}\mathrm{d}S_l$$

$$=\frac{1}{2}\int \Psi_{\underset{\smile}{ikl}}\frac{1}{2}\,e_{iklm}\mathrm{d}f_{lm}=0$$

where $\mathrm{d}f_{lm}$ is the element of surface which limits the hyper-volume, the element of which has $\mathrm{d}S_k$ as dual vector and where the surface is taken at the limit where $\Psi_{\ldots}$ vanishes.

It is then possible to use this field $\Psi_{\underset{\smile}{ikl}}$ as a *gauge* on T_{ik}, for $(T_{ik}+\Psi_{\underset{\smile}{ikl},l})$ still satisfies $(\ldots+\ldots)_{,k}=0$.

But the P_i remain completely determined even if another gauge is chosen. So, in order to determine T_{ik} completely, six relations can be assumed (the number 6 arising from the antisymmetry in $\Psi_{\underset{\smile}{ikl}}$). These are taken as the constancy of given "angular momenta"

$$M_{\underset{\smile}{ik}}(=-M_{\underset{\smile}{ki}})\equiv-\frac{i}{c}\int (x_iT_{kl}-x_kT_{il})\,\mathrm{d}S_l\,.$$

Then

$$(x_iT_{kl}-x_kT_{il})_{,l}=0.$$

[1] See A. Mercier, Principes de Mécanique analytique (Gauthier-Villars, Paris 1955) where the homogeneous formalism is elaborated completely.

Such conditions are satisfied if T_{il} is symmetric:

$$T_{il} = T_{li} = T_{\underline{il}} \, .$$

Indeed

$$\begin{aligned}(x_i T_{kl} - x_k T_{il})_{,l} &= x_{i,l} T_{kl} + x_i T_{kl,l} - x_{k,l} T_{il} - x_k T_{il,l} \\ &= T_{kl}\delta_{il} + 0 - T_{il}\delta_{kl} - 0 \\ &= T_{ki} - T_{ik} = 0 \, .\end{aligned}$$

Yet, the expression

$$T_{ik} \equiv \delta_{ik}\mathfrak{L} - q_{\alpha,i}\mathfrak{L}_{,q_{\alpha,k}}$$

is not necessarily symmetric between i and k. So it will be symmetrized by choosing conveniently $\Psi_{\breve{ikl}}$.

There is, by the way, another possibility, as we shall presently see. Since, however, these considerations are not generally covariant, let us generalize them first.

3.2.3. Action, Lagrangian Density and Energy-momentum Tensor in GRG

Let $\mathfrak{L}$ be a Lagrangian density in $\mathbb{R}^4$. The action will be defined as

$$S = \frac{1}{c}\int \mathfrak{L}\sqrt{-g}\,\mathrm{d}V$$

where the factor $1/c$ is put for reasons of dimensions.

But now, we have to assume

$$\mathrm{d}s^2 = g_{\dot{ik}}\mathrm{d}y^i\mathrm{d}y^k \, .$$

Hence, S will be a functional of the coordinates by means of the g_{ik}. We must postulate it to be an invariant. The variation δS will be different if built as a $\delta_q S$ from variations of the coordinates q or if built as a $\delta_g S$ from arbitrary variations δg_{ik} of the metric.

3.2.3.1. „Conservation“ Equations

Equations of "conservation" will be obtained from variations of the coordinates, i.e. from $\delta_q S$. Of course, these variations yield variations of the g_{ik}, but the latter, $\delta_q g_{ik}$, are no more arbitrary. They arise from infinitesimal variations imposed upon the coordinates, and these can always be regarded as infinitesimal changes of coordinates. But a change in the coordinates replaces g^{lm} by

$$g'^{ik}(y')=g^{lm}\frac{\partial y'^i}{\partial y^l}\cdot\frac{\partial y'^k}{\partial y^m}.$$

Let us therefore put

$$y'^i=y^i+\xi^i \quad \text{where the } \xi^i \text{ are infinitesimal.}$$

Then

$$g'^{ik}(y')=g^{lm}\,(\delta^i_l+\xi^i_{,l})\,(\delta^k_m+\xi^k_{,m})$$
$$\simeq g^{ik}(y)+g^{im}\xi^k_{,m}+g^{kl}\xi^i_{,l}.$$

But, on the other hand since the ξ^i are infinitesimal

$$g'^{ik}\,(y+\delta y)=g'^{ik}(y)+\frac{\partial g'^{ik}}{\partial y^l}\,\delta y^l, \quad \text{where} \quad \delta y\to\xi$$

or

$$g'^{ik}(y)\simeq g^{ik}(y)+g^{im}\xi^k_{,m}+g^{kl}\xi^i_{,l}-g^{ik}_{,l}\xi^l.$$

A detailed computation would show that this reduces to

$$g'^{ik}(y)=g^{ik}(y)+\xi^{i;k}+\xi^{k;i}.$$

Hence

$$\text{in } g'^{ik}=g^{ik}+\delta g^{ik}, \quad \delta g^{ik}=\xi^{i;k}+\xi^{k;i}.$$

In the same manner

$$\text{in } g'_{ik}=g_{ik}+\delta g_{ik}, \quad \delta g_{ik}=-(\xi_{i:k}+\xi_{k;i}).$$

Thus we can always write

$$g'_{il}g'^{lm}\simeq\delta^m_i.$$

We shall have to put $\delta_q S=0$. This $\delta_q S$ can be computed either from the $\delta_q g_{ik}$ of from the $\delta_q g^{ik}$.

First and independently from the fact that the δg_{ik} are δ_q's,

$$\delta S = \frac{1}{c}\int\left[\frac{\partial(\sqrt{-g}\,\mathfrak{L})}{\partial g^{ik}}\,\delta g^{ik} + \frac{\partial(\sqrt{-g}\,\mathfrak{L})}{\partial g^{ik}_{,l}}\,\delta g^{ik}_{,l}\right]\mathrm{d}V$$
$$= \frac{1}{c}\int\left[\frac{\partial(\sqrt{-g}\,\mathfrak{L})}{\partial g^{ik}}\,\delta g^{ik} + \frac{\partial}{\partial y^l}\left\{\frac{\partial(\sqrt{-g}\,\mathfrak{L})}{\partial g^{ik}_{,l}}\,\delta g^{ik}\right\} - \delta g^{ik}\frac{\partial}{\partial y^l}\left\{\frac{\partial(\sqrt{-g}\,\mathfrak{L})}{\partial g^{ik}_{,l}}\right\}\right]\mathrm{d}V.$$

In this expression because of Gauss's theorem, the following term vanishes:

$$\int \frac{\partial}{\partial y^l}\{\ldots\}\,\mathrm{d}V = \oint \ldots = 0$$

since the $\delta g \ldots$ must vanish at the limit on the surface of integration of $\oint$. Under these conditions

$$\delta S \sim \int\left[\frac{\partial(\sqrt{-g}\,\mathfrak{L})}{\partial g^{ik}} - \frac{\partial}{\partial y^l}\left\{\frac{\partial(\sqrt{-g}\,\mathfrak{L})}{\partial g^{ik}_{,l}}\right\}\right]\delta g^{ik}\mathrm{d}V\ .$$

This variation would take up the form

$$\delta S = -\frac{1}{2c}\int T_{ik}\delta g^{ik}\sqrt{-g}\,\mathrm{d}V$$

if

$$\frac{1}{2}\,T_{ik}\sqrt{-g} = -[\ldots]\left(=\frac{1}{2}\,T_{ki}\sqrt{-g}\right)$$

which is symmetric, since all the i, k follow from g^{ik}.

We could have made these calculations by starting with the δg_{ik}; this would have led us to T^{ik} in

$$\delta S = +\frac{1}{2c}\int T^{ik}\delta g_{ik}\sqrt{-g}\,\mathrm{d}V$$

which, by the way, follows immediately from the fact that $g^{ik}\delta g_{ik} = -g_{ik}\delta g^{ik}$.

From the δg_{ik} or the δg^{ik} found further above and from the symmetry of T^{ik} or T_{ik}, we deduce further that

$$\delta S = -\frac{1}{2c}\int T_{ik}(\xi^{i;k} + \xi^{k;i})\sqrt{-g}\,\mathrm{d}V = -\frac{1}{c}\int T_{ik}\xi^{i;k}\sqrt{-g}\,\mathrm{d}V$$

$$= -\frac{1}{c}\int \underbrace{(T^k_i\xi^i)_{;k}}_{\underbrace{\frac{1}{\sqrt{-g}}\frac{\partial(\sqrt{-g}\,T^k_i\xi^i)}{\partial y^k}}_{\oint \sqrt{-g}\,T^k_i\xi^i \mathrm{d}S_k \;=0\;(\xi^i\to 0 \text{ at the limit})}} \sqrt{-g}\,\mathrm{d}V + \frac{1}{c}\int T^k_{i;k}\xi^i\sqrt{-g}\,\mathrm{d}V$$

$$= \frac{1}{c}\int T^k_{i;k}\xi^i\sqrt{-g}\,\underbrace{\mathrm{d}V = 0}_{\text{variation principle}}.$$

Since the ξ^i are arbitrary we conclude that:

$$T^k_{i;k} = 0$$

Therefore "conservations" follow from $\delta_q S = 0$ if

$$T_{ik} = \frac{2}{\sqrt{-g}}\left[\left(\frac{\partial(\sqrt{-g}\,\mathfrak{L})}{\partial g^{ik}_{,l}}\right)_{,l} - \frac{\partial(\sqrt{-g}\,\mathfrak{L})}{\partial g^{ik}}\right].$$

But they are not true conservations! (Cf. further down 6.1, 2.)

3.2.3.2. Field Equations

The Lagrangian method with its Hamiltonian principle implies Einstein's field equations if certain hypotheses are made on S and if the metric itself is submitted to arbitrary variations. In other words, the problem consists in finding a particular (though as general as possible) action functional such that the corresponding Euler-Lagrange differential equations are found identical with Einstein's equations. To achieve this, a suitable variation has to be chosen. (Of course "conservations" will remain valid.)

For this, one separates first of all S_m (m: matter) from S_{gr} (gr: gravitational field), and puts

$$\delta S = \delta(S_m + S_{gr}) = 0.$$

In order to get S_{gr}, one needs a scalar function of the g_{ik} as well as of their first derivativs (just as in classical mechanics, the q and the $\dot{q}$ appear in $\mathfrak{L}$). We remember the existence of an invariant curvature

$$R \equiv R_i^i .$$

This might furnish our scalar. So, up to a constant, one might try

$$S_{gr} \sim \int R \sqrt{-g}\, \mathrm{d}V .$$

With regard to the domain of integration, one could imagine it such that, at its limit, it approaches a so-called *flat*, i.e. pseudo-Euclidean variety where the g_{ik} become $\pm\delta_{ik}$ and where the curvature and the Γ's tend to zero.

However, the above functional S_{gr} is not exactly what is usually introduced in Hamilton's principle because R is built up with derivatives of the g_{ik} up to the second order. But, in spite of this, one can avoid the dilemma, for it is possible to get rid of the influence of second derivatives for the following reasons:

The integrand is equal to

$$\sqrt{-g}\, R = \sqrt{-g}\, g^{ik} R_{ik} = \sqrt{-g}\, g^{ik} \{\Gamma^l_{ik,l} - \Gamma^l_{il,k} + \Gamma^l_{ik}\Gamma^m_{lm} - \Gamma^m_{il}\Gamma^l_{km}\} .$$

Now

$$\sqrt{-g}\, g^{ik}\Gamma^l_{ik,l} = (\sqrt{-g}\, g^{ik}\Gamma^l_{ik})_{,l} - \Gamma^l_{ik}(\sqrt{-g}\, g^{ik})_{,l} ,$$
$$\sqrt{-g}\, g^{ik}\Gamma^l_{il,k} = (\ldots)_{,k} - \ldots .$$

The first terms on the right hand sides of these developments yield, when put into S_{gr}, integrals of the form

$$\int (\sqrt{-g}\, A^l)_{,l}\, \mathrm{d}V = \oint \ldots = 0$$

which vanish by virtue of the limit conditions. So, the only remaining term is of the form

$$\int G \sqrt{-g}\, \mathrm{d}V$$

where G does not contain derivatives of an order superior to the first.

A further calculation would show that

$$G = g^{ik}(\Gamma^m_{il}\Gamma^l_{km} - \Gamma^l_{ik}\Gamma^m_{lm}) .$$

But, one could as well keep R, and then build

$$\delta_g S_{gr} \sim \delta \int R\sqrt{-g}\,\mathrm{d}V = \delta_g \int g^{ik} R_{ik} \sqrt{-g}\,\mathrm{d}V$$
$$= \int (R_{ik}\sqrt{-g}\,\delta g^{ik} + \delta R_{ik}\sqrt{-g}\,g^{ik} + \delta\sqrt{-g}\,g^{ik}R_{ik})\,\mathrm{d}V$$
$$= \int \left(R_{ik} - \frac{1}{2}g_{ik}R\right)\delta g^{ik}\sqrt{-g}\,\mathrm{d}V + \int g^{ik}\delta R_{ik}\sqrt{-g}\,\mathrm{d}V\,.$$

Then, only the first integral remains, whereas the second reduces to $\int (\sqrt{-g}\,A^i)_{,i}\,\mathrm{d}V \to 0$ and vanishes.

This is the variation obtained for S_{gr}.

S_m, on the other hand, can only be due to the existence of T_{ik}; hence, δS_m will be the expression

$$\delta S_m = -\frac{1}{2c}\int T_{ik}\delta g^{ik}\sqrt{-g}\,\mathrm{d}V$$

obtained further above, since T_{ik} is not to be considered as a given function of the g_{ik} but as the initial datum upon which the structure of space-time, i.e. the metric will depend (eventually by means of a curvature). Thus by producing the variations $\delta_g(S_m + S_{gr})$ they will not affect T_{ik}, and the variation principle will read

$$\delta_g(S_{gr} + S_m) = \int \left[\chi^{-1}\left(R_{ik} - \frac{1}{2}g_{ik}R\right) - T_{ik}\right]\delta g^{ik}\sqrt{-g}\,\mathrm{d}V = 0$$

where χ is a constant. Since the δg^{ik} are arbitrary, it follows that

$$S_{ik} \equiv S_{ik} - \frac{1}{2}g_{ik}R = \chi T_{ik}\,.$$

These are field equations; the constant χ is a measure of the intensity with which *matter* ($\sim T_{ik}$) "curves" the universe. The left hand side coincides with Einstein's tensor S_{ik}.

Let us insist upon the fact that the variations allowing the establishment of the field equations and those establishing the "conservation" equations are distinct. The latter resulted from variations of the coordinates which on account of the invariance of $\mathrm{d}s^2$, were $\delta_q g_{ik}$'s. They are not arbitrary variations of the g_{ik}. By producing them, one compared an "extremum" *distribution* of matter with a "varied" distribution. The variations which lead to field equations are arbitrary variations of the g_{ik}; there an "extremum" *metric* is compared with a "varied" metric.

Since the establishment of field equations shows that Einstein's tensor is proportional to T_{ik}, nothing is changed in the "conservation" equations for we know that automatically $S^i_{k;i} \equiv 0$.

This gives us the opportunity to notice that there is a formal symmetry in the relation which combines R^i_k and T^i_k. Indeed, by contracting indices and by remembering that $\delta^i_i = 4$, one obtains from the field equations

$$R^i_i - \frac{1}{2}\,\delta^i_i R = -R = \chi T^i_i \equiv \chi T\,.$$

Hence

$$\left(T_{ik} - \frac{1}{2}\,g_{ik} T\right) = \chi^{-1} R_{ik}$$

which is symmetric to

$$\left(R_{ik} - \frac{1}{2}\,g_{ik} R\right) = \chi T_{ik}\,.$$

Thus, one gets the impression that two points of view are just as good. From the first one, the distribution T^i_k of matter and energy creates the particular metric; from the other one, the metric of space-time appears through the channel of curvature R_{ik} as a distribution of matter expressed in the complex $\left(T_{ik} - \frac{1}{2}\,g_{ik} T\right)$. It is true that this second interpretation is less familiar; according to it χ^{-1} would be a kind of elasticity modulus of the universe, just as, according to Young, stresses in an elastic medium are expressed as functions of existent deformations. However, the left hand side contains the metric g_{ik} in its expression $T_{ik} - \frac{1}{2}\,g_{ik} T$, so that we should rather not consider this second interpretation as the reversal of Einstein's field equation.

Field equations in GRG, be they read in one or the other sense, are *non*-linear, contrarily to most of the field equations of physics. This means that two "fields of gravitation" both solutions of the same equations cannot be superposed in order to get a solution of the same equations again. Therefore, matter as mass, energy and momentum *is not analogous* to the electric charge and its current which, as sources, create the electromagnetic field, and as distribution is strictly conserved.

In spite of this, to summarize the results obtained above, GRG theory is not alien to the most general principle of mathematical physics: $\delta \int \mathfrak{L} \mathrm{d}V =$

$=0$. It is in that sense a dynamics, but a *mixed* dynamics, a *gravito-geometro-dynamics*. It implies the following scheme:

Field equations	→ "conservation" equations	
Variation principle with δ_g Choice of a Lagrangian	→ field equations	the choice made in the above argument implies one single constant of gravitation χ
Variation principle with δ_g without additional choice	→ "conservation" equations	
"conservation" equations	→	uniqueness of the choice of the most general conserved tensor. (Cartan's tensor $\supset$ Einstein's tensor; possibility of two interaction constants χ and λ.)

The only remaining problem is in principle of experimental nature; it consists in the determination of χ and possibly λ. In praxis, however (yet for χ only), this problem is best solved by establishing Newton's law of gravitation (or its equivalent like Poisson's equation) as an approximation of Einstein's field equations and by identifying the constant; but this is not completely satisfactory (see further down).

3.2.3.3. Trajectory of a Test-particle (Preliminaries)

By a suitable application of the variational method, still another group of equations can be found if one cares to distinguish the description of the special motion of what is called a *test-particle*[1] from the description of the totality of the universe.

For this, the action S is to be considered in $\mathbb{R}^4$, not only as a functional of the g_{ik} but also of the coordinates – z^k – of that test-particle. The z^k should not be confused with the y^k upon which the g_{ik} finally depend. This means that, on the whole, three types of variational derivatives $\frac{\delta s}{\delta \ldots}$ should be distinguished leading to three types of equations respectively.

$\frac{\delta S}{\delta_g g_{ik}}=0,$ where the field (represented by the metric) is independently varied; this principle leads to field equations.

[1] A test particle is such that the metric remains the same whether or not the test particle exists (is a world-line) in $\mathbb{R}^4$.

$\frac{\delta S}{\delta_y g_{ik}}=0,$ where the metric is varied in function of an independent variation of the coordinates y^k of space-time; this principle leads to the definition of an energy-momentum tensor and guarantees its "conservation".

$\frac{\delta S}{\delta z^k}=0,$ where independent variations of the coordinates of the test particle are produced; this yields the equation of motion and hence the trajectory of that particle.

All these results summarize and generalize at the same time those known under the name of Noether's theorem. Action is considered as constituted by additive terms, among which the one relative to the test-particle is an integral along the unknown trajectory:

$$\int \mathrm{d}V \int \mathrm{d}s \Lambda \left(g_{\dots}, \frac{\mathrm{d}s}{\mathrm{d}z}\right) \delta(y-z),$$

where

$$\mathrm{d}s=\sqrt{g_{ik}\mathrm{d}z^i\mathrm{d}z^k}$$

and where $\delta(y-z)$ is the four dimensional delta-function.

3.2.4. "Con servation" and Conservation

We have called "conservation" in quotation marks the equation

$$T^k_{i;k}=0$$

because it is not a true conservation. Indeed, the integral

$$\int T^k_i \sqrt{-g}\, \mathrm{d}S_k$$

would actually be conserved only if

$$(\sqrt{-g}\,T^k_i)_{,k}=\frac{\partial(\sqrt{-g}\,T^k_i)}{\partial y^k}=0$$

whereas

$$T^k_{i;k}=T^k_{i,k}-T^k_l\Gamma^l_{ik}=\dots$$

$$=\frac{1}{\sqrt{-g}}(\sqrt{-g}\,T^k_i)_{,k}-\frac{1}{2}g_{kl,i}T^{kl}=0.$$

The presence of the second term $-\frac{1}{2} g_{kl,i} T^{kl}$ shows that beside the energy-momentum T_i^k considered so far which is due to the matter spread throughout the universe and which produces the curvature of $\mathbb{R}^4$, the appearance of a gravitational field induces the existence of a self-energy-momentum. This field would then be its own source, which is extremely paradoxical.

There are still other difficulties.

On the one hand, any scalar density used under $\int$ in the variation principle leads to "conservation" equations. These equations are identities; this is not satisfactory, or rather, it is too satisfactory. Yet this might do.

On the other hand, the Lagrangian which leads to the true equations of conservation either coincides with the Lagrangian of the field or it does not. If it does, conservation is evident, because in that case, $\delta L/\delta_y g_{ik} = 0$ automatically. If it does not, what do the true equations of conservation have in common with the field equations? Thus, we encounter three difficulties.

In order to write down true equations of conservation, several procedures have been invented. The oldest one goes back to Einstein himself and uses a "pseudo"-tensor t_i^k of energy-momentum. (We put "pseudo" in quotation marks because it is not a pseudo-tensor in the sense of a completely antisymmetric tensor of rank 4). It is in fact possible to define t_i^k such that the sum $(T_i^k + t_i^k)$ be conserved. The "pseudo"-tensor t_i^k *does not transform* like a tensor, unless one restricts oneself to special coordinates, in particular to harmonic coordinates. This last fact has induced some authors to assume that only these coordinates are applicable to the actual description of GRG phenomena; this amounts to giving up general covariance. The expression for t_i^k is complicated; it uses the g^{ik} and the Γ_{kl}^i; but it can also be derived by means of the canonical formalism from a Lagrangian which in vacuum is equal to

$$\mathfrak{L} = \sqrt{-g}\,(\Gamma_{lm}^k \Gamma_{ki}^i - \Gamma_{li}^k \Gamma_{km}^i)\, g^{lm}$$

ifone puts

$$t_k^l \equiv \mathfrak{L}\delta_k^l - g_{mi,k} \mathfrak{L}_{,g_{mi,l}}.$$

This "pseudo"-tensor evidently vanishes in the absence of gravitational field; this is already the case if the Γ's vanish. Therefore, the non-purely gravitational energy-momentum T_i^k is conserved; hence, if this interpretation makes sense t_i^k supports the energy-momentum of a pure gravitational field but it is not covariant, for t_i^k is not a tensor.

All sorts of "pseudo"-tensors have been found: either, they are not true tensors like t_i^k, or they contain arbitrary elements which have no true physical foundation. From this, follows that the notion of gravitational energy is contingent, either because one restricts the choice of a system of coordinates to some particular class and consequently reduces, if not destroys the idea of general covariance, making of GRG a semi-special theory, or because one assumes arbitrary elements in order to guarantee true conservation and covariance simultaneously.

It seems that, strictly speeking, it is *impossible to combine the three principles* of covariance, equivalence and conservation (cf. 6.1, 2).

4. SPECIAL PROBLEMS AND EXPERIMENTAL TESTS

4.1. Typical Problems

In principle, the typical problem of GRG would consist in finding twenty magnitudes g_{ik}, T_{ik}, satisfying

1⁰ the field equations $S_m^l = \chi T^l_m$, or even C_m^l instead of S_m^l

2⁰ limit conditions still to be formulated.

If we knew the limit conditions corresponding to the actual physical state of the universe, there would be no problem left, once the equations were solved, these conditions taken into account.

But we do not know the state of the universe for our means of investigation probably do not reach its limits; moreover the problem is, in any case, so to say inextricable because of the difficulty to choose coordinates that are well adapted to the real structure of the universe.

For these reasons, particular solutions are looked for. These are either *exact* or *approached*. In order to obtain the latter, hypotheses are made about the weakness of the field intensity allowing to neglect some small terms. Exact solutions are obtained from a restricted consideration of geometrical or more generally, kinematical conditions of symmetry, the two prototypes of which are

(a) a homogeneous distribution of matter, the homogeneity referring to a 3-dimensional space as seen by observers who claim to observe a homogeneous distribution around them;

(b) a distribution which could be said to be "totally" inhomogeneous, namely such that from a center of observation, having its own world-line going at any time through that center, matter (i.e. T_i^k itself) presents a central or spherical symmetry.

In both cases (a) and (b) as well as in many other ones, a distinction can be made between static and non-static distribution; this requires a clock to be singled out, upon which temporal variation will be measured.

Static central (spherical) symmetry was the first one to be considered by Schwarzschild. It is taken as the model of isolated stars. The solution is found to split into two partial solutions: an interior one, valid in the interior of the distribution, where $T_k^i \neq 0$, and an exterior one. These two solutions must fit together. By solutions are meant the functions $g_{ik} = g_{ik}(y)$. Coordinates y^k are chosen, adapted to central symmetry. But there are several such possible choices.

The static solution has been generalized to a non-static one.

It was the totally homogeneous distribution which led to the first cosmological considerations. This is pretty clear because such a homogeneity concerns the totality of the universe whereas spherical symmetry is particularly adapted to the near surroundings of a single celestial body.

One typical problem has still to be taken into consideration. It was already known in pre-relativistic physics, i.e. strictly speaking, not within gravitodynamics but within electrodynamics because it requires the presence of a field. It is the determination of the trajectory of a test-particle. We shall begin with this last problem.

4.2. Trajectory of a Test-particle (Detailed)

A first definition of a test-particle has already been given. In GRG, a test-particle could also be said to possess so small a mass that its contribution to T_i^k can be neglected for no difference in the field is noticeable whether it is there or not.

The sun being considered as an isolated star, a planet approaching it can be assimilated to a test-particle which does not modify the

Schwarzschild solution describing its field. But the test-particle need not be a planet or a comet, it could be e.g. a photon with zero rest mass.

The trajectory of a test-particle is nothing but its world-line; but if it is to be understood as a three dimensional trajectory, a projection along a suitable time axis will have to be made; this will require the choice of some sort of clock attached to a Galilean system of coordinates in each point of the world-line (this choice will be discussed in section 4.4).

In the absence of electromagnetism, the world-line coincides with a geodesic; its equation follows from the variation principle

$$\delta \int g_{ik} \mathrm{d}z^i \mathrm{d}z^k = 0 .$$

The particular world-line will have to be fitted on to initial conditions. By introducing a parameter σ, e.g. s if the rest mass does not vanish, a Lagrangian

$$L = g_{ik} \frac{\mathrm{d}z^i}{\mathrm{d}\sigma} \frac{\mathrm{d}z^k}{\mathrm{d}\sigma}$$

can be extracted from the variational principle and the corresponding differential equations of the trajectory will read:

$$\frac{\mathrm{d}}{\mathrm{d}\sigma} \frac{\partial L}{\partial \left(\frac{\mathrm{d}z^l}{\mathrm{d}\sigma} \right)} - \frac{\partial L}{\partial z^l} = 0$$

with one of the two following accessory conditions

$\mathrm{L} = 1$ $(\sigma \equiv \mathrm{s})$ for a planet or a comet

$\mathrm{L} = 0$ for a photon.

Once the world-line is found, it has, in each of its points, a unit tangent vector u^i which, by choosing units for which $\mathrm{c} = 1$, is the quadrivelocity of the test-body. If (for $L = 1$) one puts

$$p^i = m u^i$$

with the condition

$$p_i p^i = m^2$$

and if by reversing an argument made in section 3.1 one requires that

$$p^i_{;i}=0$$

(whose interpretation is the conservation of the substance of the particle), m is nothing but the mass of the test-particle.

If matter consists of a pure fluid of (variable) density μ, but devoid of inner pressure and viscosity, its material tensor is

$$M^k_i=\mu u_i u^k,$$

and the vanishing of its divergence yields

$$\nabla_k M^k_i=(\mu u^k)_{;k}\, u_i+\mu u_{i;k}u^k=0\,.$$

Now, $u^i u_i=1$, so if we multiply this equation by u^i, and take into account the equation of continuity

$$(\mu u^k)_{;k}=0$$

which warrants the conservation of mass, we are left with

$$\mu u^i_{;k}u^k=0\,.$$

If we consider the fields u^i and μ as attached to matter contained in a small volume V, then the total mass m_0 is obtained by integrating over V

$$\int_V \sqrt{-g}\mu\delta V\equiv m_0\,.$$

Then, by Gauss' theorem, since nothing leaves the domain of integration,

$$\int(\sqrt{-g}\mu u^k)_{;k}\,\delta V=[\text{vanishing flux}]+\frac{1}{c}\int\frac{D}{Dt}(\sqrt{-g}\mu)\,\delta V$$
$$=\frac{1}{c}\frac{d}{dt}\int(\sqrt{-g}\mu)\,\delta V=0$$

which vanishes since mass is conserved. On the other hand, the following expression must also vanish, yielding, by Gauss theorem,

$$\int(\sqrt{-g}\mu u^i_{;k}u^k)\,\delta V=\{\text{vanishing flux}\}+m_0\frac{D^2z^i}{Ds^2}=0\,.$$

This describes the motion of an (extended) particle of rest mass m_0, whose center of mass of coordinates z^i lies on a time-like geodesic. Here the particle is deformable even if of small extension, but devoid of further structure or of any property other than proper mass. Thus we see that its world line is automatically a geodesic; this follows from Einstein's field equations and integrability conditions.

These examples of dust and pure fluids explain why general relativity has no use of a supplementary equation of motion beside the field equations. Several authors have extended this result in various degrees of generalization and by inventing various methods (De Donder 1924, Einstein and Grommer 1927, Einstein, Infeld and Hoffmann 1938, Fock 1939 et al.)

However, the motion of a body depends on its structure and further properties, i.e. on the form of the total tensor to be used as material tensor, thus also on the fact, that beside gravitation, other "forces" may act upon it. Such "forces" will have to be incorporated as fields into the material tensor; then the equation of motion will follow from its structure and its vanishing divergence (cf. 6.2, 3).

4.3. Central Symmetry

This is the case of a ds^2 which by means of coordinates still to be interpreted can e.g. be put into the form

$$ds^2 = d\varrho^2 + \varrho^2 f(\varrho, \tau)\,(d\theta^2 + \sin^2\theta d\varphi^2) - h(\varrho, \tau)\,d\tau^2 .$$

One interprets ϱ as a radius of action (different from a distance) counted from the center of symmetry and τ as time referred to an observer whose geodesic is the geodesic of a star bound with the center for which $\varrho = 0$. Spherical symmetry is then characterized by the *angles* θ and φ because these angles fail to appear in the functions f and h, whereas they are included in the parenthesis $(d\theta^2 + \sin^2\theta d\varphi^2)$ exactly as they do in ordinary spherical symmetry.

In the ten g_{ik}, only the two functions g and h remain as unknown.

Limit conditions are such that for any τ, f and h tend to 1 if $\varrho \to \infty$. The meaning of h depends on the definition of the clock used. As to f, it will be seen that, in case θ varies alone, a meridian circle at infinity measures indeed $2\pi\varrho$ in circumference.

For fixed values of ϱ and τ the ds^2 is that of an ordinary sphere but ϱ is not its radius. Let r be that radius. Some authors prefer to use r instead of ϱ. The corresponding substitution can be made. If ϱ(or r) and τ are variable and θ and φ fixed, another two-dimensional sub-space is obtained, in which there are lines $r = \text{const.}$ and orthogonal lines $L_\perp$. Along each line $L_\perp$ the value taken by τ at the point where $L_\perp$ crosses the geodesic of the star center can be taken as new time t. With these definitions (assuming $c=1$) the ds^2 takes over the form

$$ds^2 = A(r,t)\,dr^2 + r^2(d\theta^2 + \sin^2\theta d\varphi^2) - B(r,t)\,dt^2$$

where at the center $A(0,t)=1$, $B(0,t)=1$. This illustrates the fact that there are several ways of imposing a central or spherical symmetry, each one requiring the physical interpretation of the coordinates used.

Precisely that new form of the ds^2 had been chosen by Schwarzschild; however he had only solved the problem of the static state of a central material sphere or star, i.e for which $\frac{\partial A}{\partial t} = 0 = \frac{\partial B}{\partial t}$. In our days, variable stars must be considered, in particular pulsating ones, called "pulsars" for which τ (respectively t) appear in f and h (resp. in A and B).

By starting from the g_{ik} of the central symmetric metric, T_{ik} is found simplified: several of its components vanish; essentially T_{11} and T_{44} remain (where the index 1 corresponds to r or ϱ and the index 4 to t or τ); the other components depend on both T_{11} and T_{44}. At places where $T_{11} = 0 = T_{44}$, all the other T_{ij} vanish and, moreover, space becomes flat, so that there is in particular no gravitation field within a hollow sphere. By giving T_{11} and T_{44} as well as limit conditions it must be possible to fix completely the yet undetermined elements of the metric.

The mass M of the star is defined proportional to the following integral:

$$M \sim -\int_0^{H(t)} r^2 T_4^4 dr\,.$$

The equation

$$r = H(t)$$

describes then the variable surface of the star. It can be shown that this mass is constant in the course of time t:

$$\frac{dM}{dt} = 0\,.$$

There are two regions: an interior one for $r<H(t)$ and an exterior one for $r>H(t)$.

The *exterior field* is described by the metric

$$ds^2_{\mathrm{ext}} = -\frac{\mathrm{d}r^2}{1-\dfrac{\chi M}{4\pi r}} - r^2(\mathrm{d}\theta^2+\sin^2\theta\,\mathrm{d}\varphi^2) + \left(1-\frac{\chi M}{4\pi r}\right)\mathrm{d}t'^2$$

where

$$t' = \int\limits_0^t \exp\left[(1/2)\,\chi \int\limits_0^{H(t)} r\mathrm{e}^{\alpha}\,(T^1_1 - T^4_4)\,\mathrm{d}r\right]\mathrm{d}t\,,$$

$$\mathrm{e}^{-\alpha} = 1 - \frac{2GM}{c^2 r}\,.$$

In the course of this new "time" t' the g_{ik} appear to be static; this is the expression of a theorem of Birkhoff according to which any exterior Schwarzschild field can be transformed into a "static" field. However, only if mass is in a static state, that is if $\frac{\mathrm{d}H}{\mathrm{d}t}=0$, do t and t' coincide (up to a factor) with each other.

The interior field evidently depends upon the distribution T^i_k as a function of r (or of ϱ). By the way, the exterior field does so too since $T^1_1 - T^4_4$ appears under the integral, except in the static case when t is re-established.

A *model* is said to correspond to each choice of a distribution. Example: a perfect incompressible fluid with homogeneous density.

Schwarzschild has discovered a *singularity* named, since, after him. By examining the $\mathrm{d}s^2$, this singularity is found to arise for the value $r_S = \frac{2GM}{c^2}$ of r: the coefficient of $\mathrm{d}r^2$ jumps there from $-\infty$ to $+\infty$ and the coefficient of $\mathrm{d}t'^2$ changes its sign. As long as $r>r_s$, these coefficients tend to ± 1 for $r\to\infty$, as must be for space to be flat at infinity. But at the singularity, the coefficient of $\mathrm{d}t'^2$ is zero and that of $\mathrm{d}r^2$ tends to $-\infty$. Nevertheless, even at the singularity, the determinant $\|g_{ik}\|$ remains finite. Below the singularity, i.e. where $(r<r_s)$, the $\mathrm{d}s^2$ looses its meaning altogether. Fortunately, this singularity is at the interior of the star, thus, it concerns neither the $\mathrm{d}s^2_{\mathrm{ext}}$, nor the trajectories of test-particles (of planets and the like).

The case of the trajectory of a *test-body in the exterior Schwarzschild field* has been studied in detail. Let m_0 be its rest mass.

First case: $m_0 \neq 0$ *Planet or comet.* With the usual initial conditions for planets the geodesic found is an ellipse, the plane of which slowly moves in a way such, that the azimut of the perihelion increases.

The comparison between the slow motion of the perihelion with observation requires that χ assumes the *numerical value*

$$\chi = 2{,}07.10^{-48} \mathrm{cm}^{-1} \mathrm{g}^{-1} \mathrm{s}^{-2}.$$

Second case: $m_0 = 0$ *Photon.*

Two problems arise: On the one hand, the red shift by Doppler effect of the wave length of the photon as measured by an observer at a fixed distance from the source, whereby the photon withdraws from the source towards the observer. On the other hand, the deviation of the direction taken by the photon emitted by a far distant source, after going through the exterior field of a spherical mass.

Comparisons between observations and the theory have been made

1⁰ by calculating and observing the displacement of the perihelion of three planets (Mercury, Venus, Earth) in a field which is exterior to the sun,

2⁰ by calculating (during an eclipse of the sun) the deflection produced by the sun of the light rays coming from a distant star. Here are the results in the two cases.

Perihelion:

	Mercury	Venus	Earth	
Obs.	43″, 11 ± 0.45	8″, 4 ± 4,8	3″, 0 ± 1,2	per terrestrial century
Theory	43″, 03	8″, 3	3″, 8	

Deflection:

Obs.	between 1″, 43 and 2″, 7
Theory	1″, 75

As to the red shift of the spectral lines, it is no more considered today as a true test of GRG, because it verifies the conservation of energy and the equivalence between mass and energy rather than the field equations as such. Indeed, for this red shift, GRG yields the same result as a purely

Euclidean explanation in which the loss in frequency would be a loss in energy $\triangle(h\nu)$ arising from the necessity to overcome Newtonian attraction between the mass of the sun as a source and that of the photon $\frac{h\nu}{c^2}$ as a test particle emitted.

In former days, it has often been said that the three phenomena, motion of the perihelion, deflection of light rays and red shift, are the only experimental confirmations of GRG theory. They have even been called crucial verifications. This is, to our mind, a misunderstanding. Firstly, the third of these phenomena is not a confirmation since it concerns something else. Then, if Newton and his successors had not existed and if physics had begun with Einstein instead of Newton, Einsteinian gravitodynamics, would explain everything accounted for by Newtonian gravitodynamics, as well as the trajectory of planets including the motion of their perihelion and the form of trajectories of photons, approaching solar mass.

Indeed, as long as the metric field g_{ik} only slightly differs from

$$\begin{pmatrix} 1 & 0 & 0 & 0 \\ 0 & -1 & 0 & 0 \\ 0 & 0 & -1 & 0 \\ 0 & 0 & 0 & -1 \end{pmatrix}$$

and is in particular quasi-static, Einstein's equations reduce to the unique Poisson equation $\triangle\varphi = -4\pi G\mu$ if φ is put equal to

$$\varphi = -\frac{c^2}{2}(g_{00}-1)$$

and χ equals to

$$\chi = \frac{8\pi G}{c^4}.$$

Now, on the one hand, Poisson's equation is obtained under the assumption that inertial mass and grave mass are identical. On the other hand, the value obtained from the above expression for χ, by putting in it the known experimental value of the Newtonian gravitational constant is

$$\frac{8\pi G}{c^4} = 2{,}07.10^{-48} \text{ c.g.s.}$$

This is exactly the same as the experimental value obtained from the observation of the planets including their perihelion motion as described in Case 1 above. This, we think, is the best proof of the identity of the two masses, inertial and grave. Thus Einstein's equations appear to be a law of nature or interaction law, or, if preferred, a coupling in which the metrical field is related with (coupled to) energy-momentum distributed in the universe. Hence, all phenomena already satisfactorily explained by the Newtonian theory of gravitation are, at the same time a magnificent confirmation of Einsteinian gravito-dynamics. For this reason combined with the two cases explained above, GRG is an experimentally *very well verified* theory and not, as is often said just sufficiently "justified" by a few specific cases.

This is the opportunity to say a few words about observation in GRG.

4.4. Observation in GRG

Observation in GRG is a very tricky matter. In this section we confine ourselves to mathematical methods concerning the definition of singular coordinate systems where observation will be as little different as possible from observation as was conceived of in pre-relativistic and special relativistic physics.

The observation rests on two elementary notions already known in special relativity

1⁰) the *event* which is a point $\boldsymbol{y}$ of $\mathbb{R}^4$ where some "coincidence" takes place – a coincidence so to say of "two points" each of which would belong to the world-line of a distinct object destitute of any other dimension than precisely the length of this world-line.

2⁰) *proper time*

Every other magnitude will have to be defined explicitely or implicitely. It will then be either an invariant, in which case its observation has a unique sense, or a tensor of rank ≥ 1, in which case its observation has a relative sense. An observation is made by means of an apparatus. Such an apparatus should preferably be constituted by invariant elements. But primarily to any measuring apparatus, one device must be at hand as the primordial element which every other observation is referred to, namely an ideal clock.

The ideal clock is a world-line marked with points $\ldots, A, B, C, \ldots$ such that

$$\ldots = \int_A^B ds = \int_B^C ds = \ldots .$$

Each interval $\int_X^Y ds$ of equal length is a unit of time. By subdividing these intervals one gets arbitrarily small time intervals.

Among ideal clocks, there are such which are "manufactured" from geodesic lines. They are *geodesic clocks*. Since every mass particle doing nothing but gravitating in $\mathbb{R}^4$ moves along a geodesic, a *purely gravitational ideal clock* ("kitchen clock") is a geodesic clock.

On the contrary, an electrical charged particle coupled to the electric field e.g. in a crystal lattice (spiral in a "wrist watch") is no more a geodesic clock. Consequently, there is no reason a priori to expect that the various markings of time made on clocks of different types should coincide. By different types, we mean clocks, based on the use of bodies coupled to various fields by virtue of the various "quantum numbers" which these fields may imply according to the various interactions prevailing in nature.

Events of the first kind will be those obtained at the crossing of any world-lines with the geodesic line of a purely grave body. As purely grave bodies, one considers mainly photons, possibly neutrinos, in spite of the disadvantages attached to their vanishing rest masses, occasionaly neutrons but the latter are not strictly speaking purely grave (by the way, photons are not either purely grave because of their electromagnetic origin and of the production and anihilation of pairs; neutrinos are not either for similar reasons), or finally composite bodies, like celestial ones, such as planets, or the mass of a pendulum etc.

Actually, the only purely gravitational body or particle is a graviton. But nobody knows whether there are gravitons, and if there are, then the nonlinearity of Einstein's equations implies that gravitons can themselves exchange gravitons.

We can now move on to the measurement of spatial distances, as follows: Let L_0 be the world-line of an observer O. From the point of view of $\mathbb{R}^4$ geometry, O and L_0 are one and the same thing. But by expliciting the time t of O, one given value of that time fixes a point Q on L_0 (Fig. 1). Let P be a point of $\mathbb{R}^4$ outside L_0 on the world-line of an infinitesimal spherical mirror reflecting light in all directions. O asks what distance separates him from P. To answer this question O, at a time $t - t_1$ conveniently chosen ($t_1 > 0$) read on his ideal clock, sends a light pulse which being a photon, is identical with a nullgeodesic no. 1. The mirror in P reflects this light pulse along a null geodesic no. 2 and O

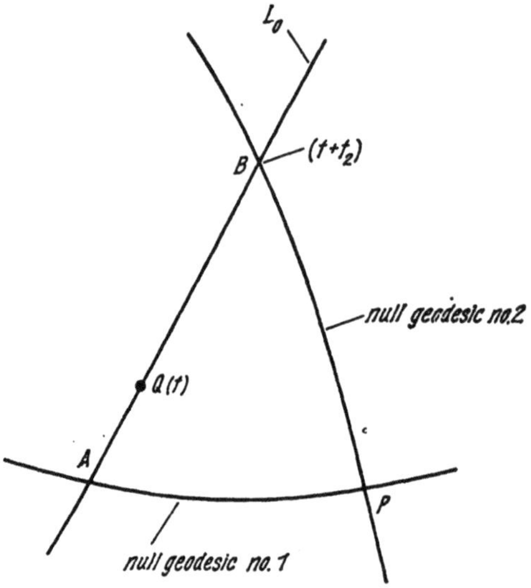

Fig. 1. Measurement of distances

receives it at time $t+t_2$ read on his ideal clock. There is one point $Q(t)$ for which

$$\Delta \equiv c\sqrt{t_1 t_2} = \sqrt{-\tau_1 \tau_2} \quad (\tau \equiv \mathrm{i}ct)$$

is maximum. At that moment Δ can be used to define uniquely the distance from L_0 to P. For this definition to be practical this distance has to be small; its quality of being small is measured by approximating sufficiently well the curved triangle ABP by an Euclidean triangle.

Before GRG was known, inertial systems were considered as privileged coordinate systems and they were defined:

1⁰) either as attached to fixed stars,

2⁰) or as left to themselves without being accelerated by anything else,

3⁰) or as systems of orthogonal families of straight lines used as Cartesian coordinates.

There is no analogy for this last image in GRG because a $\mathbb{R}^4$ allowing for the choice of Cartesian coordinates is automatically flat. What can be done, is to define locally an inertial system whose time coordinate coincides with the time of the ideal clock of the observer, that is, with his graduated world-line.

Moreover, there is something analogous to

$$\frac{\mathrm{d}^2 r}{\mathrm{d}t^2} = 0$$

by means of what Walker calls ***Fermi's displacement***. Let $y^k = y^k(s)$ be a world-line γ and

$$u^k \equiv \frac{\mathrm{d}y^k}{\mathrm{d}s}$$

the unit vector tangent to γ. We shall assume that all along γ, $u^k u_k > 0$ (time like orientation of γ). Successive absolute derivatives of u^k can be considered, whence the series

$$u^k,\, v^k = \frac{Du^k}{\mathrm{d}s}, \quad w^k = \frac{D^2 u^k}{\mathrm{d}s^2}, \quad \mathfrak{w}^k = \frac{D^3 u^k}{\mathrm{d}s^3}, \ldots .$$

From $u^k u_k = 1$ we get

$$u^k \frac{Du_k}{\mathrm{d}s} = 0 = u_k \frac{Du^k}{\mathrm{d}s}.$$

Hence

$$\boldsymbol{u} \perp \boldsymbol{v},$$

whereas $\boldsymbol{w}$, $\mathfrak{w}$ . . . are perpendicular neither to $\boldsymbol{u}$ nor to $\boldsymbol{v}$, nor to one another.

By normalizing them suitably, one can take the first four of these vectors as a local basis to build an orthogonal tetrad

$$A^k \equiv u^k,\, B^k,\, C^k,\, D^k$$

which satisfies the following conditions

$$\frac{DA^k}{\mathrm{d}s} = bB^k, \quad \frac{DB^k}{\mathrm{d}s} = cC^k + bA^k,$$

$$\frac{DC^k}{\mathrm{d}s} = \mathrm{d}D^k - cB^k, \quad \frac{DD^k}{\mathrm{d}s} = -\mathrm{d}C^k.$$

The constants b, c, d, can be determined by requesting that

$$B^k B_k = 1, \quad C^k C_k = 1, \quad D^k D_k = 1.$$

The following relations can be verified:

$$A^k B_k = 0,$$

$$A^k C_k = A^k \frac{DB_k}{ds} - bA^k A_k = \frac{1}{b} A^k \frac{D^2 A_k}{ds} - b$$
$$= \frac{1}{b} \frac{D}{ds}\left(A \frac{DA_k}{ds}\right) - \frac{1}{b} \frac{DA^k}{ds} \cdot \frac{DA_k}{ds} - b$$
$$= -\frac{1}{b} b^2 B^k B_k - b$$
$$= 0$$

etc. All vectors are orthonormal.

Given a (contravariant) vector V^k, it is called "Fermi displaced" along γ if in the course of this displacement,

$$\frac{DV^k}{ds} = A^k V_i B^i .$$

Then

$$\frac{D}{ds}(V^k A_k) = -V_k \frac{DA^k}{ds} + A_k \frac{DV^k}{ds} = -V_k \frac{DA^k}{ds} + V_i \frac{DA^i}{ds} = 0 .$$

If V^k is successively made equal to A^k, B^k, C^k, D^k, the Fermi displacement leaves the tetrad in identically the same relation to the world-line γ, since the changes undergone by the four podes hcppen in the direction of γ. The tetrad can thus be used as a local inertial system of coordinates and by following it, we get the analogous relation to $\ddot{\vec{r}} = 0$.

The Fermi displacement can be generalized to the Walker displacement for which

$$\frac{DV^k}{ds} = bV_i (A^k B^i - A^i B^k)$$

and which leaves orthogonal the tetrad.

In this way, (quasi-) inertial systems can be defined in GRG.

4.5. Equivalence „Principle" and Eddingtonian Numbers

Originally, Einstein had enunciated a "principle" of equivalence. Equivalence was meant as the identification of inertial and grave mass. Yet, for us, this is no more a principle but a statement to be used as the definition of either the Newtonian potential entering Poisson's equation or of the gravitational force by means of the negative gradient of the same potential.

However, there are several steps in the establishment of this statement (of this "principle").

First step: statement of a weak principle: the trajectory of a (purely grave) test-particle depends neither on the mass, nor on the nature of this particle. This statement allows to define the force of gravitation acting on the particle in the Newtonian approximation.

Second step: observational statement: in every point of $\mathbb{R}^4$, there exists a class of local inertial systems of coordinates with regard to which the above test-particle eventually found in that point moves inertially. These systems of coordinates are "laboratories falling freely". For them, space-time is locally flat.

Third step: invariability statement from point to point throughout the universe. Before enunciating the statement, let us notice how difficult it is to compare standards of measurement in different points of $\mathbb{R}^4$. In gravitodynamics, three standards are needed. In old days, since the decimal system had been introduced c.g.s. were used. Suppose now that current theories include so-called universal constants. If three of them are dimensionally independent, any system of dimensions like c. g. s. can be derived from them. But not any constant is suited for such a reduction. We posses at least two kinematico-mechanical constants, c and h (or $\hbar$) and although $\hbar$ stems from quantum theory we can use it. The constanats e (elementary charge) and k (Boltzmann's constant) for instance stand outside the frame of gravitodynamics and wherever we look for it we can never find the third necessary constant. If we did possess a universal length which would have to be, not the result of a computation based on some interaction (in particular on gravitation with G or χ), but a theoretical primordial element different from a constant of interaction, the set would be at hand. A universal period or a universal mass would do too. But it would be wrong to count an interaction constant like χ among the universal natural units because interaction constants are precisely those which are to be numerically determined by experiment as functions of the chosen units.

Let M_0 be a universal mass still to be found – perhaps the mass of a proto-particle from which the rest masses of all other particles would be calculated by means of an appropriate theory according to which they might be resonances of the proto-particle. With $\hbar$, c, and M_0 for instance, no pure number can be obtained by products of their powers. But in gravitodynamics if a further interaction constant is added, then a pure number can be obtained, namely

$$\gamma \equiv \frac{\hbar^4 \chi}{c^3 M_0^3}.$$

The best known of such pure numbers is however, not of gravitodynamic, but of electrodynamic nature (within special relativistic quantum theory). It does not imply a mass but a charge, it is the well known fine structure constant discovered by Sommerfeld:

$$\alpha \equiv \frac{e^2}{4\pi\varepsilon_0 \hbar c} \simeq \frac{1}{137}.$$

It would seem extraordinary if pure numbers, even possessing a fundamental physical meaning like α or γ, should vary from point to point in the universe, in other words if $\alpha = \alpha(y)$, $\gamma = \gamma(y)$, ...

The third step concerning the statement of the principle of equivalence would now consist in saying that firstly, the planicity spoken of in the second step is the same in every point of $\mathbb{R}^4$ – which means that the physical laws are the same in all freely falling laboratories – and secondly, that each dimensionless physical constant is the same pure number when measured in different points of $\mathbb{R}^4$ by means of apparatuses attached to laboratories falling freely.

It is clear that, beside $\alpha, \gamma, \ldots$ other pure numbers can be constructed as soon as one possesses specific informations about the elements of the world. For example, the ratio of two masses is a pure number; it could be the ratio between the mass of the proton and the mass of another particle which might or might not be a resonance of the proton etc. But it seems that only those numbers implying interaction constants are really important. They are sometimes called *Eddingtonian numbers* because this author established a list for the first time.

Here is the list which can be established nowadays. The numbers are classified according to an increasing numerical value:

Since we have eliminated from GRG interactions differing from gravitation, it would be best to ignore all such numbers built from ε_0, g or the constant of strong interaction. The only remaining number within the first order of magnitude is the cohesion constant of the universe

$$\frac{GM}{Rc^2} = 0{,}3$$

	Eddingtonian number	order of magnitude of its numerical value
1st order of magnitude	constant of strong interaction	10
	cohesion constant of the universe	$\frac{GM}{Rc^2} \approx 0{,}3$
	fine structure constant of electro-magnetic interaction	$\alpha \equiv \frac{e^2}{4\pi\varepsilon_0 \hbar c} \approx 1/137$
2nd order of magnitude	constant of weak interaction	$\left(\frac{g m_\pi^2 c}{\hbar^3}\right)^2 \approx 3.10^{-14}$
3rd order of magnitude	$\frac{\text{gravit. interaction}}{\text{el. magn. interaction}}$ (per H-atom)	$\chi c^4 m_e m_p \varepsilon_0 / 2e^2 = G m_e m_p 4\pi\varepsilon_0 / e^2 \approx 5 \cdot 10^{-40}$
	$\frac{\text{atomic period}}{\text{age of the universe}}$	$\frac{e^2/\pi\varepsilon_0 m_0 c^3}{T} \approx 0{,}2 \cdot 10^{-49}$
4th order of magnitude	inverse of the number of baryons in the universe	$\frac{m_p}{\frac{4}{3}\pi R^3 \varrho} \approx 3 \cdot 10^{-79}$

Notations

m: mass $\begin{cases} e: \text{electron} \\ p: \text{proton} \\ \pi: \text{pion} \end{cases}$

G: Newtonian constant } of gravitation
χ: Einsteinian constant }

g: coupling constant of beta-radioactivity $\approx 1{,}4 \cdot 10^{-49}$ erg. cm^3

M: total mass of the universe

ϱ: average mass density in the universe $\approx 10^{-30}$g $\cdot$ cm^{-3}

T: age of the universe $\approx 10^{10}$ years

R: radius of the universe $= cT$ (Hubble) $\approx 10^{28}$ cm and in the fourth order of magnitude the reciprocal number of particles contained in the universe

$$\frac{m_p}{\frac{4}{3}\pi c^3 T^3 \varrho} \approx 3 \cdot 10^{-79}.$$

These two numbers include $R = cT$ in the denominator. Now, R, proportional to T increases with the age of the universe. Consequently, if these two Eddingtonian numbers are to remain constant numbers, then, in the first order of magnitude, G or χ must vary unless M or c do so and in the fourth order of magnitude g, m_p or c must vary. An increase of M would amount to the creation of matter.

Dirac and a few others have speculated about the variability of factors in the above numbers and constants drawing for instance the conclusion that G (or χ) ought to diminish in the course of time for the two constants of the third order to remain simultaneously of the same order of magnitude (since the second one is proportional to $1/T$), under the assumption that M_e, m_p, ε_0 and i_e (and c) are invariable. But then the variation of G would be the inverse of what is suggested by the first order of magnitude. By the way, $\frac{\varepsilon_0}{T}$ might as well remain constant; in that case, no variation of G would be necessary.

Moreover, the cohesion constant cannot be a "constant" in the same sense as the other Edingtonian numbers; it is rather the present measure, fortuitously found of the order of magnitude of unity, of an influence of the universe in its totality upon the order of magnitude $O(g_{ik})$ of the metric tensor g_{ik} in each point of $\mathbb{R}^4$. This would amount to a local influence of distant origin or better of global origin. This is the kind of thought implied by a principle stated by Mach according to which we could write either

$$O(g_{ik}) \approx \int_0^R \frac{\varrho \mathrm{d}^3 V G}{r c^2} \approx \frac{GM}{Rc^2}$$

or

$$G^{-1} \approx \int_0^R \frac{\varrho \mathrm{d}^3 V}{r c^2} \approx \frac{M}{Rc^2}.$$

From that point of view, if the order of magnitude $O(g_{ik}) \approx GM/Rc^2$ of the g_{ik} is due to matter in the universe as a whole, the influence of the nearby sources is added, for instance the influence of the sun in its immediate neighbourhood creating a perturbation of the order of m_0/rc^2 at a distance r, or that of a galaxy in its interior etc. This would explain why g_{ik} is not constant but fluctuates around the value $O(g_{ik})$, by and by when approaching, or moving away from, individual masses.

Let us not forget at this point, that contrarily to a Newtonian conception (or rather to Poisson's view of a scalar potential φ of gravitation) which by the way, does not propagate at finite velocity but "acts at a distance", the g_{ik} build a tensor: gravitation is liable to a symmetric tensorial field of rank 2.

Although it is a local theory–owing either to the restriction to local measurements or to the local nature of Einstein's equations–GRG, as suggested by the remark about Mach's conception, is compatible with a statement of global nature; this compatibility is invoked more and more frequently in the construction of models of the whole universe. Meanwhile, the propagation of the metric field implies the question of the existence and eventually of the meaning of gravitational waves; this question itself is bound with the possibility of gravitational signals or news (is there something called gravitational radiation?) travelling from region to even further region of the universe. Moreover, the type of geometry may depend on various circumstances and consequently, the universe as a whole will have properties allowing for a really cosmological interpretation of Nature. Two big problems are thus put. The first one concerns gravitational radiation and will probably have to be submitted to a careful and long analysis before another still bigger problem can be verily tackled, we mean the quantifyability of the gravitation field. (Let us recall that quantum theory is issued from the quantization of electromagnetic radiation alone known in those days.) The second big problem is the cosmological one in the sense on a theory of the entire universe and of the observation of connected predictable effects (cf. 5.1, 2).

4.6. Gravitational „Waves“

Here again, we must put one notion in quotion marks. For, there are gravitational waves only at some approximation, in contradistinction to classical electrodynamics where electromagnetic waves arise as soon as charges are accelerated. Indeed, the existence of electromagnetic waves follows from an exact solution of Maxwell's equations for regions void of electric charge. Gravitational waves do not follow from the general Einstein equations if e.g. T_i^k is put equal to zero in analogy to $s^i=0$ (s^i: quadridensity of electric current in electrodynamics). What can be shown, is that a sufficiently weak gravitational field behaves in a way similar to what a field does if consisting of waves. But, strictly speaking "sufficiently weak" implies that the field tends toward zero, hence, there is no field at the wave-consistent limit.

Assume that the g_{ik} differ practically very little from Galilean g_{ik} which read

$$(\overset{\circ}{g}_{ik}) = \begin{pmatrix} 1 & 0 & 0 & 0 \\ 0 & -1 & 0 & 0 \\ 0 & 0 & -1 & 0 \\ 0 & 0 & 0 & -1 \end{pmatrix}.$$

that is,

$$g_{ik} = \overset{\circ}{g}_{ik} + h_{ik}, \quad \left|\frac{h_{ik}}{g_{ik}}\right| \ll 1, \quad h_{ik} = h_{ki}.$$

The Γ^i_{kl} will be small and curvature reduce to

$$R_{iklm} = \frac{1}{2}\,(h_{im,k,l} + h_{kl,i,m} - h_{km,i,l} - h_{il,k,m}),$$

$$R_{ik} = \frac{1}{2}\,(-\overset{\circ}{g}{}^{lm} h_{ik,lm} + h^l_{i,kl} + h^l_{k,il} - h^j_{j,ik}).$$

Let us notice that if these formulae are valid for one system of coordinates, they will have to be valid also for another one obtained by an infinitesimal transformation. We can then impose four conditions upon h_{ik} as follows

$$\left(h_i^k - \frac{1}{2}\,\delta_i^k h_j^j\right)_{,k} = 0.$$

The h_{ik} satisfying them can therefore be gauged by means of

$$y^i \to y'^i + \xi^i \quad \text{and} \quad h_{ik} \to h'_{ik} = h_{ik} - \frac{\partial \xi_i}{\partial y^k} - \frac{\partial \xi_k}{\partial y^i} \quad (= h'_{ki})$$

as long as

$$\Box \xi^i = 0 \quad \left(\Box \equiv \overset{\circ}{g}{}^{lm} \frac{\partial^2}{\partial y^l \partial y^m} = \frac{\partial^2}{c^2 \partial t^2} - \Delta\right).$$

By calculation, curvature is found equal to

$$R_{ik} = -\frac{1}{2}\,\Box h_{ik}$$

whence, in vacuum, where $T_{ik}=0$,

$$\Box g_{ik}=\Box h_{ik}=0 \quad \text{and} \quad \Box \Gamma^i_{kl}=0\,.$$

The conditions imposed imply that in a plain gravitational wave h_{ik} reduces to two non-vanishing components contained in the plane perpendicular to the direction of propagation.

Thus there would be gravitational waves under the assumption of solutions with weak intensity of the field equations. One does not know whether there are waves as exact solutions without such an assumption for, in order to know, one would have to produce, from Einstein's equations which are nonlinear, solutions resembling in vacuum at least asymptotically spherical waves at a great distance of the sources and going over without discontinuity into the solutions valid in the interior of matter distribution. This problem is nearly inextricable.

Going back to weak waves, since they seem to exist, let us remark that if in quantum field theory differential equations are looked for corresponding to particles of spin 2 without rest mass $\left(\text{analogous to Dirac's equation for mass particles of spin } \frac{1}{2}\right)$ they are found to assume the form

$$\Box \underset{\circ}{\varphi}{}_i^k=0$$

with the accessory condition

$$\underset{\circ}{\varphi}{}_{i,k}^{\,k}=0$$

where the index o under φ specifies the reference to a Minkowski space $\mathbb{E}^4$, for in that theory one restricts covariants to Lorentz transformations of special relativity. Apart from the difference between $\mathbb{E}^4$ and $\mathbb{R}^4$, these equations are the same as those which predict weak gravitational waves in vacuum. The conclusion has been drawn that, if ever gravitation were also to be a quantum phenomenon like electromagnetism in the case of photons which are the quanta of the electromagnetic field, there ought to exist "quanta of the gravitational field". They are called *gravitons*. Gravitons would have a zero rest mass and a spin 2.

One does not know, up to this day whether there are really gravitational waves transporting energy and whether consequently it makes sense to talk about the quanta of gravitational radiation. Experimental results about these waves are still controversial, though some people claim to have caught gravitational pulses by suitably built devices. At least his claim says that the pulses obtained could be neither light pulses, nor the result of seisms.

The non-linear equations leave little hope. Moreover, since there should not be any privileged coordinate system in GRG, and since this does not allow to construct uniquely an energy-momentum tensor, even if there were waves, T_i^k would vanish in vacuum since vacuum is defined by $T_i^k = 0$; consequently, even if these waves did transport energy, this very energy-momentum transport should not be noticeable!

From this, as well as from the difficulties concerning energy conservation, several authors, Fock and others, drew the conclusion that *coordinate conditions* perhaps must be imposed on GRG, in particular the harmonicity condition, the idea of which goes back to de Donder (1921) and to Lanczos (1922). This condition is analogous to the one proposed by Lorentz in classical electrodynamics. In electrodynamics, Lorentz's condition is made possible by gauge invariance, and gauge invariance is intimately connected with the existence of electric charge. It is a way of linearizing the theory and making it accessible to Fourier's analysis.

But, conditions like that destroy some of the most characteristic properties of GRG (see 6.4).

4.7. Cosmology

Cosmology in GRG can be said to be at the same time a cosmogeny. This depends on the point of view taken.

The cosmogenic tendency is to propose a model of the universe for which a *universal time* is formally separated from a remaining 3-dimensional space in order to show in the view of man and at human scale how the world evolves. Let T designate that universal time. Proposed models have been made upon the assumption that the metric reads:

$$d\sigma^2 = dT^2 - d\sigma^2$$

in which $d\sigma^2$ does not contain dT. Only six components of the metric are still to be given.

Moreover, further assumptions are made about the expression of $d\sigma^2$. A first distinction is made between static and non static universes. In static universes, the six remaining components of g_{ik} do not depend on T.

Of course, one never knows a priori whether an assumed metric will yield a correct description of the universe.

Another tendency consists in looking for a suitable transcription of the idea that the universe is filled up with incoherent matter, with a kind of dust distributed homogeneously, i.e. with constant density. This last assumption needs further precision: what does homogeneous mean? A particle of the dust has a world-line γ; one assumes that whatever be the value of T defining a point P on γ, and whatever be the particular world-line γ, the 3-dimensional density at point P is measured by numerical magnitudes which may depend on T only.

4.7.1. Cosmological Models

4.7.1.1. Einstein's Model

R being a positive constant, one puts

$$ds^2 = dT^2 - R^2 [d\omega^2 + \sin^2\omega(d\theta^2 + \sin^2\theta \, d\varphi^2)]$$
$$(0 \leqq \omega \leqq \pi, \quad 0 \leqq \theta \leqq \pi, \quad 0 \leqq \varphi \leqq 2\pi) .$$

From this, the energy-momentum tensor can be computed starting with Cartan's tensor in the field equations:

$$T_{ik} = \varrho \delta_i^0 \delta_k^0$$

where the index 0 is chosen in order to make $T = y^0$ and where $\varrho = \text{const} >$ > 0 is a mass density. The cosmological constant λ assumes then the value

$$\lambda = \frac{1}{R^2} = \frac{\chi \varrho}{2} .$$

For $T = \text{const}$, the three-dimensional sub-space is closed (hyperspherical), hence the universe appears in this perspective to be a cylinder: topological product of a hyper-sphere by an open straight line (of time) but there are still other topological decompositions and the enumeration of the spatio-temporal interpretations of the model is so far not yet exhausted.

4.7.1.2. De Sitter's Static Model

This model replaces the closed three-dimensional sub-space of Einstein by an open one. However, it is *empty* (otherwise the total mass contained would be infinite), which means that the (uniform) density of matter tends everywhere towards zero.

4.7.1.3. Friedmann's Properly Cosmogonic Model

This model is not static; it generalizes the preceding ones, using a metric of the form

$$ds^2 = dT^2 - R^2(T) d\sigma^2,$$

where several assumptions can be made on $d\sigma^2$. In particular, one gets

either a closed three-dimensional subspace of curvature $\varepsilon = +1$,
or a flat three-dimensional subspace of curvature $\varepsilon = 0$,
or an open three-dimensional subspace of curvature $\varepsilon = -1$.

Again

$$T_{ik} = \varrho(T)\, \delta_i^0 \delta_k^0,$$

but where $\varrho(T)$ is no more a constant. Nevertheless, the mass of the universe

$$M \equiv \frac{4\pi}{3} \varrho R^3 = \text{const} > 0$$

is found constant. The radius R of the 3-dimensional section of that universe varies according to Friedmann's equation

$$\dot{R}^2 = \frac{\chi M}{4\pi R} \frac{1}{3} \lambda R^2 - \varepsilon, \quad \dot{R} \equiv \frac{dR}{dT}$$

which is of elliptic type.

This model furnishes the explanation of an *expansion* of the universe. It is commonly considered as the explanation of Hubble's effect, which is a red shift of the spectral lines emitted by gallactic sources; the further away the source from the observer, the greater the red shift is.

One succeeds too in observing the density of matter, at least of the matter, the existence of which has been verified. (Neutrinos remain practically invisible; their density may be considerable.) Thus the density of galaxies was estimated equal to

$$\varrho \text{ (galaxies)} \approx 6 \cdot 10^{-29}\, \text{g} \cdot \text{cm}^{-3} \text{(G. Field 1959)}.$$

For the velocity of recession calculated under the assumption that Hubble's effect is a Doppler-effect referred to the distance of the source, the estimate is

$$\frac{\dot{R}}{R} \approx 10^2 \mathrm{km} \cdot \mathrm{s}^{-1} (\mathrm{Mpc})^{-1}.$$

Moreover, it is not found exactly constant; a relative acceleration of the order of magnitude

$$\frac{\ddot{R}}{R} \approx -\left(\frac{\dot{R}}{R}\right)^2$$

seems superposed. From these values, it should be possible to draw the (contemporary) values of the cosmological constant λ and of the radius R of the universe and consequently the mass M and the age T of the universe.

Actually the deceleration of the expansion is weak, the expansion itself $\frac{\dot{R}}{R}$ is, in first approximation a constant, namely,

$$\frac{\dot{R}}{R} \approx \frac{R}{T} \cdot \frac{1}{R} = \frac{1}{T}.$$

Hence, the observation of Hubble's effect does not, in first approximation, yield more than the age of the universe. The latter has been found

T of the order of magnitude 10^9 to 10^{10} years.

Not having R, one has not M either. Only the density remains as estimated above. By the way, Friedmann's equation contains ε and λ which have to be determined. Observations do not allow to do it very precisely. Present observations suggest that the universe is closed and λ practically zero.

Finally one cannot consider Hubble's effect and the estimates drawn from it as a serious experimental verification of Einstein's theory of GRG, not only because observations lack precision but also because nothing allows to claim that Friedmann's model is *the* correct model of the universe. Nevertheless, the expansion predicted for that model remains a very strong argument of qualitative nature, for until now no other model outside the scope of Einstein's Principles predict it.

4.7.2. Other Models

Following Goedel's instigation, other models have been suggested; they are not a priori cosmogenic. They allow the existence of a non-infinitesimal density but combine it with lateral motions of mass escaping an observation of the type of Hubble's effect.

4.8. Gravitational Collapse

By gravitational collapse, is meant disappearance of mass during which, in a classical sense the gravitational attraction surpasses the pression of the massive fluid and the whole mass precipitates towards a denser configuration in a sort of *implosion* (the opposite of an explosion) accompanied by a corresponding liberation of energy. It has been possible to elaborate an elementary theory of the collapse in the case of an initial distribution with spherical 3-dimensional (spatial) symmetry. The easiest is to start with the Schwarzschild metric.

The g_{ik} of Schwarzschild's solution are often expressed with the help of exponentials, as follows:

$$ds^2 = -e^{\lambda} dr^2 - r^2 (d\theta^2 + \sin^2\theta d\varphi^2) + e^{\nu} dt^2$$

where

$$\lambda = \lambda(r, t), \quad \nu = \nu(r, t)$$

or, by using so-called isotropic coordinates $\bar{r}, \theta, \varphi, t$,

$$ds^2 = -e^{\mu} (d\bar{r}^2 + \bar{r}^2 d\theta^2 + \bar{r}^2 \sin^2\theta d\varphi^2) + e^{\nu} dt^2$$

where $\mu = \mu(\bar{r}, t)$, $\nu = \nu(\bar{r}, t)$.

The new $\bar{r}$ is interpreted as the distance really measured from the origin.

By restricting himself to the case where ν and λ (or μ) do not depend on t assuming a perfect fluid of constant density, Tolman has studied the relation between pressure p, density ϱ and the coordinates. Considering that pressure is the trace of the 3-dimensional tensor of stresses composing a part of T_i^k, and that density is its term with indices 44, it is found, by taking pressure and density as measured at rest with regard to the fluid that

$$\frac{dp}{dr}+\frac{1}{2}(\varrho+p)\frac{d\nu}{dr}=0.$$

This Tolman equation has been generalized to arbitrary densities by Oppenheimer and Volkoff, restricting themselves to the case of static spherically symmetric configurations of a so-called cold mass (i.e. where temperature does not figure). Wheeler has extablished it as a consequence of a variation principle where mass is the varied quantity; in other words, it is an equilibrium state equation under given conditions of mass and extension.

Let us first consider a constant density.

The exterior Schwarzchild solution yields

$$e^{\nu}=1-\frac{2m}{r}-\frac{\Lambda r^2}{3}=e^{-\lambda}$$

where m is the mass and Λ the cosmological constant both expressed in units for which $\chi=8\pi$ (Tolman's notations). The interior solution can be obtained by integration and by adapting it to the exterior solution at the surface H. This yields

$$\text{in the interior}\quad\begin{cases} e^{-\lambda}=1-\dfrac{r^2(\Lambda+8\pi\varrho)}{3}=1-\dfrac{r^2}{R^2}\\ e^{\nu/2}=A^*-B\sqrt{1-\dfrac{r^2}{R^2}}\end{cases}$$

The pressure is then found to be

$$p=\frac{1}{8\pi R^2}\left(\frac{3B\sqrt{1-r^2/R^2}-A^*}{A^*-B\sqrt{1-r^2/R^2}}\right)+\frac{\Lambda}{8\pi}$$

where

$$R^2=\frac{3}{\Lambda+8\pi\varrho}.$$

If one neglects Λ in the expression of p (which is reasonable as long as one does not tackle problems of cosmic dimensions),

$$A^*=\frac{3}{2}\sqrt{1-\frac{H^2}{R^2}},\quad B=\frac{1}{2}.$$

The mass can be obtained from the radius H of the sphere; in the chosen units:

$$M = \frac{4\pi}{3} H^3 .$$

Now pressure cannot be imaginary. Therefore, for the solution to make sense, we must assume

$$H^2 < R^2$$

i.e.

$$H^2 < \frac{3}{4\pi\varrho}$$

whence

$$M < \frac{R}{2}.$$

These relations show that there is a limit to the possible extension of such an ideal fluid sphere of given constant density as well as a limit to its mass when the radius is known.

If these limits are exceeded either because the "star" is too big or because it is too massive, which is finally the same on the ground of above relations, there is no static solution; still more, there is no equilibrium between the pressure which tends to expand the sphere and its own gravitational attraction which tends to contract it; a gravitational implosion must follow which is called collapse. This phenomenon is not purely Einsteinian; Emden had already in former days, determined the equilibrium between pressure and gravitation on the basis of the Newtonian law of gravitation in his theory of polytropic gases. The Tolman-Oppenheimer-Volkoff equation is thus the one to be taken as the starting point for all polytropic models or for any generalization of these within Einsteinian gravito-dynamics. Of course, in such a generalization the hypothesis of a constant density will have to be dropped.

If we draw on a graph, as a function of r the factor $e^{\nu/2}$ (with an asymptote $= 1$) equal to the ratio between the proper time interval ds and the interval dt of the time coordinate, several curves are found for various masses (Fig. 2), two domains for possible masses M_{possible}, and impossible masses $M_{\text{impossible}}$, are separated by a critical mass curve M_{critical}.

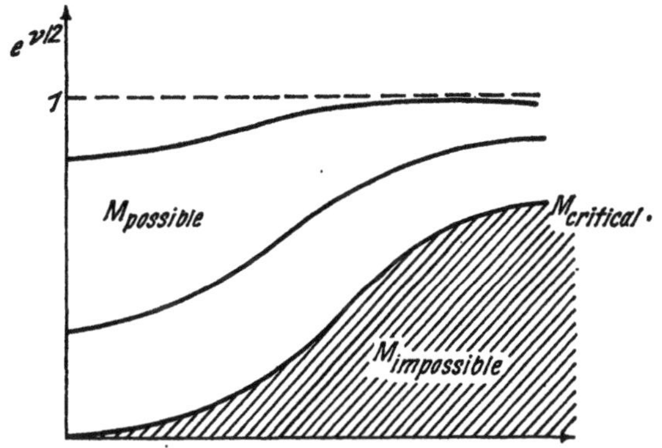

Fig. 2. Possible and impossible masses of a star:
Latent energy as a function of distance from its center

The advantage of this graph is that $e^{\nu/2}$ measures the latent energy at correspondent distance since it is analogous to the gravitational potential in Emden's equation. This is seen by computing the injection energy μdA necessary to bring dA baryons from infinity up to a distance r from the center, for it can be shown then (Wheeler et al.) that

$$\mu(r)^{\nu/2} = \text{const.}$$

In order to go through with the analysis of these spheres, a generalization of Emden's polytropes is necessary either by assuming $\varrho = \varrho(p)$ or $\varrho = \varrho(r)$ or $p = p(r)$.
Zel'dovich for instance, has assumed

$$\varrho = \text{const}/r^2 \quad \text{for} \quad 0 < r \leqq H$$

whence

$$m(r) = \int_0^r \varrho \pi r^2 dr = \text{const } r.$$

Consequently $Z \equiv m(r)/r = \text{const}$ and it can be shown that the total mass-energy of the sphere is equal to

$$M = \text{const } A^{2/3} (1 - Z)^{1/3} Z^{1/2},$$

where A is the given number of baryons entering the sphere. By varying Z from 0 to 1, a maximum $M_{\max}$ is found for M at $Z = 3/5$, whereas the energy M is weak at the two extremes $Z \simeq 0$, $Z \simeq 1$. But $Z \simeq 0$ means that

$\mathbb{R}^4$ is weakly curved, $Z \simeq 1$, that $\mathbb{R}^4$ is strongly curved around the center of symmetry, the latter case corresponding to a high density. This case for a given H leads precisely to collapse. This shows that there is an energy barrier of the order of magnitude M_{max} which must be overcome for a configuration to go from an equilibrium state of weak density over to a collapsing configuration.

Wheeler has developed an argument with the purpose to show that even in more or less any case generalizing the polytropes or Zel'dovich's model, a similar barrier will have to be overcome and this barrier will remain finite even if it becomes very big. One can then imagine that if a star evolves from the height of this barrier which it had somehow reached and falls down nearly to nothing by collapsing, it looses a trmendous amount of energy nearly equal to the height of the barrier.

Moreover, for a given radius, the pressure at the center is a function of mass; it reaches infinity for finite mass, radius and total number of baryons (Misner and Zapolski). Hence, there is a critical mass which, estimated as a function of log ϱ_0, shows a curve resembling damped oscillations (Fig. 3).

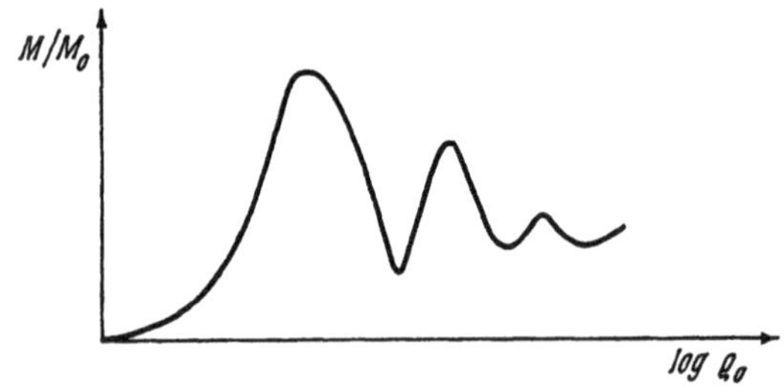

Fig. 3. Mass versus logarithm of the central density

The portions of the curve descending from left to right describe unstable equilibria subject to collapse. By drawing ϱ_0 as a function of the radius H of the massive sphere, a similar graph is obtained. The top points on these oscillations are critical points, the existence of which had been announced by Landau.

Let us add, that if the considerations referring to the collapse had first been drawn with reference to the particular Schwarzschild metric, they can be generalized and they have even been found compatible with Friedmann's metric.

Hence, there are two scales: a collapse at Schwarzschild scale of star-like spherical masses, and a collapse at a universal scale. If the latter took place, we should be destroyed.

It seems that at least the collapse at Schwarzschild scale happens, for local liberation of huge amounts of energy has been observed, and no other explanation has, so far, been given.[1]

5. EINSTEIN'S PRINCIPLES OF EQUIVALENCE

5.1. The Principles of Weak and Strong Equivalence and the Experimentes of Eötvös

The empirical basis of Einstein's general relativity is the so-called principle of weak equivalence. Eötvös has tested the universal proportionality between inertial and passive grave mass, and Dicke continued this test up to an exactitude of $1 : 10^{-11}$. From this very proportionality, Einstein was coming to the equivalence between the effects of gravitation and those of inertia in the general dynamics of particles and fields.

In order to discuss the possible generalizations of the general theory of gravitation as Einstein proposed it in 1915, a distinction must be made between weak and strong equivalence. It rests primarily on the fact that only weak equivalence can be established upon observation.

Mass enters Newtonian gravito-dynamics according to three senses: firstly as the bearer of inertial resistance, m_I; secondly as a constant telling how strongly a given gravitational field is coupled with a body upon which this field is acting,–this is what we shall call passive gravitational mass m_P; thirdly as a characteristic constant of the source of the field of gravitation; this is what we shall call active gravitational mass m_A.

Newton's equation of gravitational motion uses the two first concepts:

$$m_I \frac{d^2 x^\nu}{dt^2} = -m_P \frac{\partial \varphi}{\partial x_\nu}, \quad \nu = 1, 2, 3.$$

The third is involved in the Newtonian potential

$$\varphi = -\frac{m_A}{r}.$$

[1] More about the question of "collapses" and "black holes" and its physical meaning see 5.6 and 7.3.4.

Thus, two senses of mass refer to the behaviour of a body B in a given gravitational field of components $F_\nu = -\partial_\nu \varphi$,

$$m_I \frac{d^2 x^\nu}{dt^2} = m_P F^\nu \equiv K^\nu.$$

We may calculate from this the inertial resistance $m_I = \dfrac{K^\nu}{\dfrac{d^2 x^\nu}{dt^2}}$ and the passive grave mass m_P appears as a coupling constant which couples the body B into the external gravitational field.

On the other hand, the active grave mass m_A determines the influence of the body B upon the gravitational field. Hence, it has no bearing upon the motion in a *given* field of gravitation. It solely tells how strong the body B gravitationally influences another body B'.

Eötvös' tests teach in the first instance that inertial and passive grave mass are equal:

$$m_I = m_P .$$

If, accordingly, all bodies are said to behave in the same manner in a given gravitational field independently of their size and physical or chemical nature, i.e. to be equally accelerated, this proposition is just the principle of weak equivalence.

This principle says nothing in regard to how the gravitational influence of B or B' depends upon the nature of B or B' and/or upon their particular situation or state.

The denial of such a dependence follows, within Newtonian mechanics, from the principle of action and reaction, according to which in the interaction (mutual coupling) of any two bodies $\overset{(1)}{B}$ and $\overset{(2)}{B}$, either

$$\overset{(1)}{m}_I \frac{d^2 \overset{(1)}{x}{}^\nu}{dt^2} = -\overset{(2)}{m}_I \frac{d^2 \overset{(2)}{x}{}^\nu}{dt^2} \quad \text{or} \quad \overset{(1)}{m}_P \underset{(1)}{\partial_\nu} \varphi = -\overset{(2)}{m}_P \underset{(2)}{\partial_\nu} \varphi ,$$

these two equations being equivalent on the ground of the principle of weak equivalence.

Hence, Newtonian mechanics implies Eötvös' theorem according to which the ratio of active and passive gravitational mass to inertial mass of one and the same body is a constant independent of that particular body

$$\frac{m_A}{m_I} = f = \text{const.}$$

This universal constant must also be geometrical invariant. It can artifically be made equal to unity, if suitable units are chosen. The proposition asserting that passive grave mass, active grave mass and inertial mass are–up to a dimensional factor depending upon the units system–equal, is called principle of strong equivalence:

$$m_I = m_P = m_A \quad (f=1)\,.$$

5.2. The Principle of Weak Equivalence in GRG

The principle of inertia in the theory of special relativity

$$\frac{d^2x^i}{ds^2}=0, \quad i=0,1,2,3\,,$$

is valid in this particular form only if Cartesian coordinates are used. The equation is therefore invariant only with regard to Lorentz-transformations changing from a coordinate system Σ over to another one $\bar{\Sigma}$ in uniform translation with regard to Σ. The threefold infinity of coordinate-systems, which are obtained from one particular system in which the principle of inertia is valid by all possible Lorentz-transformations, constitutes the set of all inertial systems.

The passage from an inertial system Σ to a non-inertial one Σ which is accelerated towards Σ lets appear supplementary inertial forces, for which neither pre-relativistic, nor special relativistic mechanics can provide a physical cause.

This is why the special theory of relativity has to be primarily restricted to the use of inertial coordinate systems–however, in the frame of the formalism of this theory, changing over to a non-inertial system from an inertial one is nothing but a change of coordinates, whereby the new spatial coordinates x'^ν depend upon time: $x'^\nu = x'^\nu(\underline{x}, t)$. Their specific dependence upon t is determined by the particular law of acceleration. This is why Einstein assumed that in a generally formulated theory of relativity, where all the new coordinates depend upon all the old ones, the physical laws are the same in all coordinates-systems. This is what is called "general covariance" with regard to arbitrary charges of coordinates

$$y'^i = y'^i(\underline{y}), \quad y \equiv \{y^l\}\,.$$

Laws as formulated according to general covariance are per definition valid also in non-inertial systems; hence there will be no supplementary so-called "inertial forces" when changing to non-inertial systems.

The generally covariant form of the inertial principle valid for special relativity is, as we have learnt, given by the equation of the geodesic (3.2, 3.3)

$$\frac{d^2y^i}{ds^2}+\Gamma^i_{kl}\frac{dy^k}{ds}\frac{dy^l}{ds}=0\ ,$$

and the general equation of motion will be of the form

$$m\left(\frac{d^2y^i}{ds^2}+\Gamma^i_{kl}\frac{dy^k}{ds}\frac{dy^l}{ds}\right)=K^i.$$

In the latter equation, the term

$$m\Gamma^i_{kl}\frac{dy^k}{ds}\frac{dy^l}{ds}$$

describes the particular force of inertia arising from the particular choice of coordinates. By choosing inertial coordinates, it is made to vanish. The change from inertial coordinates x^l to non-inertial ones y^n is given by

$$\Gamma^m_{nl}=\frac{\partial^2x^a}{\partial y^n\partial y^l}\frac{\partial y^m}{\partial x^a}\ .$$

The equivalence of grave and inertial mass means that an observer attached to a coordinate-system who studies the motion of a mass-point cannot decide, whether the acceleration of this mass-point is due to gravity or to an inertial force. This indiscrimination between inertia and gravity is only locally valid. It does not allow to show, that gravitation must be described by the same means as inertial forces in accelerated coordinate-systems, viz. by means of a metric different from that of the Minkowskian space.

Hence, the line-element of Minkowskian space or its square

$$ds^2=\eta_{ik}dx^idx^k$$

characterizes space free of gravity; $(x^0, x^1, x^2, x^3)\equiv(x, y, z, t)$ constitutes inertial system.

In a gravitational field, the square of the line-element reads

$$ds^2=g_{ik}dy^idy^k\ ,$$

and the metric tensor g_{ik} describes the field. In each point $\underline{P}$ of space, a coordinate-system can be chosen such that

$$g_{ik}=\eta_{ik}+\frac{1}{2}\, g_{ik,rs}\,(x^r-x^r_{\underline{P}})\,(x^s-x^s_{\underline{P}})+O_3\,. \tag{1}$$

This is a generalization of the principle of equivalence, for it describes the field in a local inertial system; at point $\underline{P}$, there are no forces. One can interprete $\underline{P}$ as the center of gravity of a box falling freely at a given time. If it is possible to introduce a coordinate-system for which the line-element is everywhere Minkowskian, then space is free of gravitational forces and the forces appearing when other (non-inertial) coordinate-systems are used, are purely inertial forces. Thus globally, it is possible to distinguish between inertia and gravity.

Equivalence of inertial and grave mass then gets its right formulation within a Riemannian manifold. This is done by introducing first at every point of space-time local Minkowskian coordinates representing a local inertial system. A particle not submitted to forces moves in an inertial system along a straight line. The invariant properties of straight lines–autoparallelism or extremalty–must also be observable on the corresponding trajectories in a general coordinate-system. Extremalty, it will be noticed, is a simpler property, because it deals merely with the metric: From

$$\int ds\equiv\int\sqrt{g_{ik}\frac{dy^i}{ds}\frac{dy^k}{ds}}=\text{extremum}$$

follows

$$\frac{d}{ds}\left(g_{ik}\frac{dy^k}{ds}\right)-\frac{1}{2}\,g_{kl,i}\frac{dy^k}{ds}\frac{dy^l}{ds}=0\Leftrightarrow u^i_{;l}u^l=0\,,$$

whereas autoparallelism requires the definition of a displacement. If the displacement is not produced by means of the Christoffel symbols for metric, or parallel displacement, autoparallelism may yield a curve different from the extremal one. Thus only extremalty unambiguously defines the result looked for. The relativistic expression of the equivalence between grave and inertial massis therefore are to be found in the principle of the geodesics: mass-points free of forces move in a Riemannian space, i.e. in a gravitational field, along geodesics of the metric of that space.

This suggests, that there exists a dynamical identity for the material energy-momentum-tensor in the form (see 3.2)

$$T^{ik}{}_{;k} \equiv T^{ik}{}_{,k} + \Gamma^i_{kl} T^{kl} + \Gamma^k_{lk} T^{il} = 0 ,$$

which is again a still more general expression for the principle of equivalence. However, the principle does not say anything about the equations governing the metric g_{ik}. The vanishing of the covariant derivative of the matter-tensor amounts to saying that locally the energy-momentum theorem applies.

In the theory of special relativity, this latter theorem follows from field equations; there, the field is of electromagnetic nature. The next generalization of the principle of equivalence therefore consists in assuming the local validity of the special relativistic field equations for the non-gravitational fields.

Einstein's principle of weak equivalence reads: gravitational forces are undistinguishable from forces of inertia in a worldpoint $\underline{P}$ (event). The components of both corresponding fields are determined by Christoffel's symbols

$$\left\{ {i \atop kl} \right\} = \frac{1}{2} g^{ir} (-g_{kl,r} + g_{lr,k} + g_{rk,l}) = \Gamma^i_{kl} .$$

In other words, the fundamental metric tensor g_{ik} is a potential for the field of gravitation as well as for the field of inertia: the distinction between gravitation and inertia is at any point arbitrary and depends merely upon the choice of the system of coordinates.

If the coordinate-system is chosen at $\underline{P}$ in a way such that forces of inertia and gravitational forces compensate each other, we can write equation (1) at $\underline{P}$

$$g_{ik}|_P = \eta_{ik}, \quad (\partial_l g_{ik})|_P = 0 . \tag{2}$$

Hence according to the theory of general relativity, the theory of special relativity applies within infinitesimal domains of space-time.

If there are no fields beside gravitational and inertial ones, the principle of equivalence further implies that there exists a coordinate-system, in which the possible trajectories, issued at a time t_0 from a point P_0: x_0^ν ($\nu = 1, 2, 3$) of three-dimensional space of a mass-point B, are straight lines along which this body moves unaccelerated. In this system–called Riemannian system of normal coordinates–all bodies B

projected from a point $P_0\{t_0, x_0^\nu\}$ are submitted to the principle of inertia. In other words, the Riemannian system of normal coordinates is an inertial system with regard to P_0.

If furthermore the principle of inertia applies for all other bodies susceptible to be projected from any point at any time, space-time possesses a Minkowskian structure: There is no gravitation anywhere. Reciprocally, we can say that the existence of gravitation is ascertained by the fact that there is at least one finite position of space-time in which the structure is non-Minkowskian but generally Riemannian. The mathematical condition for this to be the case is that Riemann's tensor of curvature does not vanish.

General relativity theory is also to be considered as a generalization of the special theory of relativity on the ground that, in it, the acceleration of a body has lost its absolute significance. Indeed, by the use of a suitable Riemannian system of normal coordinates, this acceleration is made to vanish. What retains an absolute significance is, however, the relative acceleration between two bodies. For, if two bodies are at infinitesimal distance from each other when moving exclusively under the influence of a gravitational field, their relative acceleration amounts to

$$\left(\Delta \frac{dx^\nu}{ds}\right)_{;l} \frac{dx^l}{ds} = R^\nu_{abn} \frac{dx^a}{ds} \frac{dx^b}{ds} \Delta x^n$$

where Δx^n is their distance; now the left-hand side of this equality is the afore-said relative acceleration; the right-hand side is a tensor; but a tensor vanishes in a coordinate-system only if it vanishes in all coordinate-systems; hence the right-hand side has a meaning independently from the particular coordinate-system at use. It has e.g. the following significance: Let us consider several mass-points, which produce a negligeable field of gravitation; let these mass-points be in relative rest at time $t=0$. If they are submitted during a time interval Δt to a gravitational field possessing a Riemannian tensor R^i_{klm} and after Δt left to themselves, they will move with relative velocities among one another which are given by the time integral $t=0$ till $t=\Delta t$ of the right-hand side of above relation. Thus, according to H. Weyl and M. v. Laue, the relation from special to general relativity can be formulated like this: The principle of special relativity assumes the relativity of velocities and the absoluteness of accelerations (and of velocity differences); the principle of general relativity assumes the relativity of accelerations and the absoluteness of differences in acceleration (of the "deviations"[1]); from the latter follows in particular the absoluteness of rotation.

By construing an inertial system of coordinates in $\underline{P}$, all the equations of special relativity are valid with respect to it even if a gravitational field is at hand. The condition for this, however, is that these equations withhold their canonical character, meaning that they do not involve higher derivatives than the first ones.

Now, Einstein's principle of general covariance requires that all equations of physics remain invariant in form during changes of coordinates. This has been achieved by writing them as tensorial equations. This implies a far reaching feature: if an equation like that is valid at each point $\underline{P}$ in a coordinate-system, it is valid at each point P in any other coordinate-system. Since the transformation rule of all terms in such an equation is known from the transformation rule of the coordinates, it suffices to write down the equation in the particular system, e.g. in a Riemannian system of normal coordinates; its general form is then in principle known for any other system.

Keeping this in mind, Einstein's principle of equivalence requires then that by writing in a generally covariant form the equations of special relativity theory, the influence of the gravitational field upon the behaviour of physical systems described by these equations is automatically taken into account. Hence, on the ground of the principle of general relativity, the coupling of the gravitational field onto all other physical fields (or bodies) is unequivocally determined. The covariant formulation of canonically described systems, as borrowed from specially relativistic equations for non-gravitational fields, can be considered as a rule of construction of the action of gravity upon these fields; this, surely, is the climax of abstraction in the evaluation of the implications of the principle of weak equivalence (cf. next section).

But beside this weak principle, Einstein's theory of gravitation also assumes *strong equivalence* as it is postulated in the following proposition of Eötvös:

In a laboratory falling freely, i.e. in a coordinate-system endowed

[1] A well-known case of deviation are the tides: The gravitational field of the Sun, resp. of the Moon, is not measurable on the free moving Earth; what is measurable is the difference of these fields between different points of the surface of the Earth.

with geodesic metric (2), at a point $\underline{P}$, all gravitational effects vanish; this assumes that, unaware of the existence of external gravitational fields, the active gravitative of bodies inside the laboratory remains.

5.3. Equivalence and the Principle of Relativity

The structure of space-time of general relativity allows for three groups of transformations, which coincide in the case of special relativity and can then be identified. Those are:

a) The *Einsteinian group* of general coordinate transformations

$$y \to y'^k = y'^k(\{y^i\}) ,$$

b) the Lie group of possible motions of bodies or modifications of fields which happen under no forces and without deformation of the structure of these fields within space-time.

In the flat space-time of special relativity, every real displacement of a body referred to a given coordinate-system can be mathematically replaced by a change (translations) of the coordinate-system keeping the body in a fixed position. The condition for this to be also the case in a non-flat Riemannian space-time manifold at least for infinitesimal changes, is given by Killing's equation (see 6.2)

$$\xi_{i;l} + \xi_{l;i} = 0 .$$

This condition is not always fulfilled, for it has no solution for the most general Riemannian manifold.

c) The Lorentz-group describing the exchange of reference systems. A reference system is to be thought of as realized by the aggregate of $(3+1)$-dimensional normal rods and clocks.

Mathematically, these aggregates are represented by the 4×4 components of a field of tetrads h_i^A $(A=0, 1, 2, 3;\ i=0, 1, 2, 3)$.

The theory of relativity assumes that all measurements must be made on invariants only. Such invariant measures are either properly physical invariants (like electric charge, entropy, pressure, or action function), of the result of the projection of tensorial components $\tau^{ik\cdots}$ upon a chosen reference system itself represented by the tetrad field $h_i^A(\underline{y})$. In the latter case, orthonormality conditions are assumed in the form

$$\begin{aligned} &h_i^A h_k^B \eta_{AB} = g_{ik}\,, \\ &h_A^i h_B^k g_{ik} = \eta_{AB}\,, \\ &h_A^i \equiv g^{ik} \eta_{AB} h_k^B\,, \end{aligned} \tag{1}$$

where g_{ik} (and g^{ik}) designates the metric of space-time and η_{AB} the Minkowskian diagonal matrix

$$(\eta_{AB}) = \begin{pmatrix} 1 & 0 & 0 & 0 \\ 0 & -1 & 0 & 0 \\ 0 & 0 & -1 & 0 \\ 0 & 0 & 0 & -1 \end{pmatrix}.$$

Magnitudes of the form

$$\tau^{AB\ldots} = h_i^A \ldots h_k^B \tau^{ik\ldots}$$

are then invariant in the sense of scalars (a scale is applicable to them), for they are invariant with regards to coordinate transformations $y'^i = y'^i(\{y^l\})$.

From (C) equ. (1), the metric $g_{ik}^{(y)}$ appears uniquely determined by the tetrads h_i^A, whereas the metric of a variety V^4 determines only the tetrad field of the homogeneous Lorentz-group.

Tetrads can be Lorentz-transformed:

$$h_i^A = \omega^A{}_B h_i^B, \quad \text{with} \quad \omega^A{}_B \omega_C{}^B = \delta_C^A; \tag{2}$$

these new tetrads still satisfy equ. (1).

In general, the matrices $\omega^A{}_B$ of these Lorentz-transformations are arbitrary functions of the coordinates y^l of V^4, which are only submitted to continuity and derivability conditions: $\omega^A{}_B = \omega^A{}_B(\{y^l\})$. These Lorentz-transformations are not operations in space-time, they are "rotations" in the local Minkowskian spaces tangent to V^4 at the respective points $\underline{P}$. Hence, equ. (2) is not a coordinate-transformation and has nothing to do with the Einstein-group.

All tensors and geometric entities in V^4 are Lorentz invariant as long as no indices of tensors enter the description of properties of space-time V^4. It is the case in the normal Riemannian geometry of general relativity. This includes the phenomenological representation of matter which requires regular tensors.

Physically speaking, the tetrads yield in every point $\underline{P}$ the local reference system Σ_P, which is e.g. realized by giving three special relativistically defined rods of measure and one special relativistically defined standard clock. Local Lorentz-invariance of physical and geometrical entities means here that the relative orientation of the tetrad-fields at various points $\overset{(i)}{\underline{P}}$ of V^4 is of no relevance and that the corresponding reference systems $\Sigma_{\underset{P}{(i)}}$ can be in a state of arbitrary motion with respect to one another (principle of general covariance). Thus, the projection of tensor components into the tetrad-field is nothing but the comparison of the standard rods or clocks with the magnitudes under measurement. These projections are the natural components of the space-time magnitudes $\tau^{ik\ldots}$ referred to the given measuring apparatures. The observed natural components transform as Lorentz-tensorial components in the course of the transformations of the Lorentz-group (b), e.g.

$$\bar{\tau}^A = \omega^A{}_B h_i^B \tau^i = \omega^A{}_B \tau^B .$$

Instead of giving the 16 components $h_i^A(\underline{y})$ of the tetrad-field, it is possible to use the equivalent set of metric spin vectors $\sigma_i^{\mu\dot{\nu}}(\underline{y})$ which we shall encounter in the next section. However, then, the group of Lorentz-transformations is to be replaced by the group of unimodular spin-transformations.

Before going over to its consideration, let us examine the bearing of the principle of equivalence upon these groups. As a matter of fact, only the first and the third groups (a) and (c) are involved, for the second one (b) describes rather the internal structure of quite particular fields of gravitation.

The principles of equivalence can now be reformulated. For tensor-fields, only the Einstein-group is of importance, since all tensorial expressions are themselves Lorentz invariant. So for them, the principle reads as follows: As long as the dynamical equations for "tensorial matter" are canonical in character (no derivative higher than first ones), the influence of gravitation upon the dynamics of that matter is obtained by writing these equations explicitly according to Einsteinian covariance. (They are in any case Lorentz-invariant.)

The formulation of Einsteinian covariance requires the introduction of the metric tensor and its first derivatives (viz. in the Christoffel symbols). But since the first derivatives are of purely affine character, the principle of equivalence also says, that locally, i.e. within the infinitesimal vicinity of a point $\underline{P}$ of space-time, the equations of special relativity are valid even in the presence of a gravitational field.

For "spinorial matter" this "older" expression of Einstein's principle of weak equivalence is insufficient, because the equations valid in the frame of special relativity for that spinorial matter are automatically invariant with regard to the Einstein-group. Tensorial fields must first be obtained by fusion of spinorial fields. Then the principle of weak equivalence reads in its newer form: The influence of the gravitational field upon the dynamics of spinorial matter is obtained by formulating the canonical equations of its motion both in a Lorentz-covariant and an Einsteinian covariant manner. This newer expression contains the older one, but the converse is not true.

Yet, the ten g_{ik} and their derivatives are not enough to elaborate the Lorentz-covariant spinorial fields. In order to do it, one needs all the 16 tetrad components or the 16 components of the four spin-vectors beside their derivatives. This explains the use of a symbol with three indices, $\sigma^{A\mu\dot{\nu}}$, which is a vector with regard to the Einsteinian group but not a Lorentz-covariant magnitude; it can be made to vanish locally by a suitable Lorentz-transformation.

5.4. Covariant Spinor Calculus

There are two equivalent, though dual, representations of tensorial fields which fulfil the general principle of relativity:

a) The Lorentz-covariant representation, which uses magnitudes $\Phi^{A\ldots}_{B\ldots}$ themselves invariant with regard to coordinate transformations;
b) The coordinate covariant representation, which uses Lorentz invariant magnitudes $\Phi^{i\ldots}_{k\ldots}$.

Their equivalence is founded on the uniqueness of the correspondences

$$\Phi^{A\ldots}_{B\ldots} = h^A_i \ldots h^k_B \ldots \Phi^{i\ldots}_{k\ldots},$$

$$\Phi^{i\ldots}_{k\ldots} = h^i_A \ldots h^B_k \ldots \Phi^{A\ldots}_{B\ldots}.$$

Such a unique correspondence does not apply to spinorial magnitudes. For, the introduction of spinors implies the replacement of the univocal tensorial representation of the Lorentz-group by a bi-univocal unimodular representation. To the tetrads of the reference system correspond metric spin vectors according to

$$\sigma^{l\mu\dot{\nu}} = h^l{}_A \sigma^{A\mu\dot{\nu}} .$$

In this relation, $\sigma^{A\mu\dot{\nu}}$ is a constant Pauli-matrix; the Greek indices μ and ν are spinor-indices and take the values 1 and 2; the dot on $\dot{\nu}$ is to distinguish one index from the other.

Orthogonality reads here by means of γ-matrices: From tensorial magnitudes, we get spinorial magnitudes like this:

$$\Phi_{\mu\dot{\nu}} = \sigma_{A\mu\dot{\nu}} \Phi^A = \sigma_{l\mu\dot{\nu}} h^l{}_A \Phi^A = \sigma_{l\mu\dot{\nu}} \Phi^l .$$

The ordinary differential

$$\mathrm{d}\Psi_\nu \equiv \Psi_{\nu,l} \mathrm{d}y^l$$

of a spinor is not a spinor, since

$$\bar{\Psi}_{\nu,l} = (\alpha_\nu^\mu \Psi_\mu)_{,l} = \alpha_\nu^\mu \Psi_{\mu,l} + \alpha_{\nu,l}^\mu \Psi_\mu .$$

If we want to define a spinor from the differential, we must introduce a compensation by means of a spinorial affinity (or affine connection) $\Lambda^\alpha_{\beta l}$; it transforms according to

$$\bar{\Lambda}^\alpha_{\beta l} = \alpha^\mu_\beta \alpha^\alpha_\nu \Lambda^\nu_{\mu l} + \alpha^\alpha_\lambda \alpha^\lambda_{\beta,l} .$$

The covariant (i.e. Lorentz-covariant) derivative of a spinor is then defined by

$$\Psi_{\nu\|l} \mathrm{d}x^l \equiv (\Psi_{\nu,l} - \Lambda^\mu_{\nu l} \psi_\mu) \, \mathrm{d}x^l .$$

The spinorial affine connection is at the same time to be a space-time tensor, else the Lorentz-covariant derivative of a spinor would not be coordinate covariant. The only affinity of this kind which can be built from the spin-vectors and their derivatives has been found by Infeld and van der Waerden equal to

$$\Lambda^\alpha_{\beta l} = \frac{1}{2} \sigma^{i\alpha\dot{\nu}} \sigma_{i\beta\dot{\nu};l} = \frac{1}{2} \sigma^{i\alpha}_{\dot{\nu};l} \sigma^{\dot{\nu}}_{i\beta} .$$

There are then also mixed tensorial and spinorial magnitudes like

$$\Phi^k_{\nu|||l} = \Phi^k_{\nu,l} + \Gamma^k_{rl} \Phi^r_\nu - \Lambda^\alpha_{\nu l} \Phi^k_\alpha .$$

This latter expression yields the generally covariant constancy of the metric spin-vectors

$$\sigma^{k\mu\dot{\nu}}_{\|\|l} = \sigma^{k\mu\dot{\nu}}_{,l} + \sigma^{r\mu\dot{\nu}}\Gamma^{k}_{rl} + \sigma^{k\alpha\dot{\nu}}\Lambda^{\mu}_{\alpha l} + \sigma^{k\mu\dot{\alpha}}\Lambda^{\dot{\nu}}_{\dot{\alpha}l} = 0 \,.$$

This can be seen by computing the covariant derivative like this:

$$\sigma^{k\mu\dot{\nu}}_{\|\|l} = \sigma^{k\mu\dot{\nu}}_{,l} + \sigma^{r\mu\dot{\nu}}\Delta^{k}_{rl}$$

which vanishes, since Δ^{k}_{rl} is the integrable Einsteinian affinity (2.15, 3.5)

$$\Delta^{k}_{rl} \equiv h^{k}_{A} h^{A}_{r,l} = \sigma^{k\mu\dot{\nu}}\sigma_{r\mu\dot{\nu},l} \,.$$

With these relatives at hand, the covariant derivative of the γ-matrices is found equal to zero:

$$\begin{aligned}\gamma_{\alpha\beta\|\|l} = \gamma_{\alpha\beta\|l} &= -\gamma_{\alpha\lambda}\Lambda^{\lambda}_{\beta l} - \gamma_{\lambda\beta}\Lambda^{\lambda}_{\alpha l} \\ &= -\Lambda_{\alpha\beta\lambda} + \Lambda_{\beta\alpha\lambda} = 0 \,.\end{aligned}$$

Hence, as must, for epistemological reasons, be, the equations involving spinorial fields can be made coordinate invariant as well as—according to the principle of relativity—Lorentz-covariant. However, if spinors of odd rank $2n+1$ were used, no Lorentz-invariance or coordinate-covariance would be at hand.

For this reason, general relativity will have to require that all equations of physics should be made Lorentz-covariant and coordiante-invariant. If one restricts the the consideration to tensorial fields, the dual condition can be alternatively postulated, namely, that there equations be made Lorentz-invariant and coordinate-covariant.

In the theory of special relativity, the metric spin-vectors (resp. the tetrad-fields) will be conventionally fixed in a way such, that by fixing them one defines a pseudo-Euclidean distant parallelism. This distant parallelism distinguishes the inertial systems among all possible reference systems. In these inertial systems, in case pseudo-Cartesian coordinates be used, the tetrads assume the special form

$$h^{A}_{l} = \delta^{A}_{l}$$

and the metric spin-vectors coincide with the Dirac-matrices.

As far as a theory of gravitation admits essentially of Lorentz-covariance, the fields of tetrads remain free; in such a case, the principle of weak equivalence for "spinorial matter" can be expressed in an Einsteinian form like for "tensorial matter", viz.: The metric g_{ik} and the tetrads h_i^A can be locally chosen as to ensure the coincidence between the equations of motion of that spinorial matter within a gravitational field, and the corresponding equations of motion in special relativity.

5.5. The Principle of Strong Equivalence

The principle of strong equivalence follows from that of weak equivalence in the above formulation when the latter is completed by the remark that the influence of gravitation upon tensorial matter is uniquely determined by giving the metric g_{ik}. The structure of space-time being, then, completely determined by the metric, all physical processes including those issued from gravitation admit of the fact that the special theory of relativity is valid in the infinitesimal vicinity of any world point.

The difference between special and general relativity consists there in the fact that, for special relativity, the infinitesimal neighbourhoods of world-points are all in the same Minkowskian space-time manifold, whereas, for general relativity, each such neighbourhood belays to another Minkowskian manifold. The proposition, according to which special relativity is valid in each neighbourhood even if a gravitational field exists, is the expression of the principle of strong equivalence.

From that principle follows reciprocally, that space-time geometry and hence also the gravitational field are exclusively and completely determined by the metric tensor g_{ik}. If for gravitation further field magnitudes–e.g. the constant f of gravitation conceived as a field-function, or the 16 components h_i^A, or $\sigma_i^{\mu\nu}$ ... – are imposed, then the principle of strong equivalence is broken, for the transformations available for the Einstein group are insufficient, if all manifestations of gravity are to be locally eliminated.

For the dynamics of tensorial matter, strong and weak equivalence imply the same. For this reason, breaking strong equivalence for tensorial matter (contrary to fulfilling weak equivalence) can only be noticed from the gravitational influence of that matter. This means that then the universal proportionality between active and passive gravitational mass is broken, viz. that instead of being a universal constant, the factor of "proportionality" becomes a function which may explicitly or implicitly depend upon the field. But then, the principle of action and reaction is also broken: breaking strong equivalence breaks the action-reaction-law.

For spinorial matter there is a difference, both kinematical and dynamical, between the implications of weak and of strong equivalence. Only strong equivalence requires Lorentz-covariance of the whole theory, i.e. of the dynamics of matter plus the field equations for gravitation. Weak equivalence, on the other hand, allows to fix the tetrad-field from the gravitational field up to rigid Lorentz-rotations.

Whereas there is no difference between weak and strong equivalence as far as tensorial fields are concerned, the dynamics of spinorial fields depends essentially upon which principle is assumed valid: of weak or strong equivalence, because if only weak equivalence is valid, a space-time manifold can be parallelized at distance (up to rigid Lorentz-rotations) within specialized reference systems, and such specialized reference systems enter the equations for the spinorial fields—whereas tensorial entities obtained from the fusion of spinorial ones are again Lorentz-invariant and depend solely upon the g_{ik} and their derivatives.

Therefore, if one adopts the viewpoint according to which only tensorial entities are directly observable, there should be no difference between the implications of weak or strong equivalence as to the observed local dynamical behaviour of matter; only gravitational interactions—which transcend locality—can themselves lead to a distinction between the two equivalences.

5.6. Einstein's Gravitodynamics as Theory of Strong Equivalence

Einstein developed in 1915 the theory of gravitation based on the principle of strong equivalence. It says that g_{ik} alone determines both geometry and gravitation. The field equations for gravitation must be Einstein-covariant differential equations in which the homogeneous part is to contain only the g_{ik} and their derivatives. The condition that these equations yield the old Newtonian theory of gravitation as an approximative led to the following general expression

$$R_{ik} + a g_{ik} R + \lambda g_{ik} = -\varkappa T_{ik} \tag{1}$$

where a and λ are constants, and $\varkappa \equiv \frac{8\pi G}{c^4}$, G Newtonian constant of gravitation. These relations surely fulfil the principle of strong equivalence, since the Ricci-tensor R_{ik} and the curvature scalar $R=g^{ik}R_{ik}$ are construed from g_{ik} and its first and second derivatives only.

General Einstein-covariance implies that if $g_{kl}(y)$ is a solution of (1), all Einstein-transformed metrics

$$g'_{mn}(y') \equiv \frac{\partial y^k}{\partial y'^m} \frac{\partial y^l}{\partial y'^n} g_{kl} \tag{2}$$

are solutions, too. Consequently, there must be four identities between the ten components of the metric. (They are mathematically simple consequences of the Bianchi-identities for the Riemann-tensor of curvature.) Then

$$R^k_{i;k} = \frac{1}{2} R_{,i} \Leftrightarrow -\varkappa T^k_{i;k} = \left(\frac{1}{2} + a\right) R_{,i} \,. \tag{3}$$

Now, if the field equations (1) are also to be compatible with the principle of weak equivalence, without leading to overdetermination of the field, the latter relations (3) must also be consequences of the gravitodynamics of tensorial matter, which follow from weak equivalence. But from the Einstein-covariance principle, the special relativistic theorem of conservation

$$\partial_k T^k_i = 0$$

goes over to a generally covariant theorem of the form

$$T^k_{i;k} = 0 \,.$$

Comparison with (3) yields the value

$$a = -\frac{1}{2}$$

for the constant a.

Whence the Einsteinian gravitation equations:

$$C_{ik} \equiv R_{ik} - \frac{1}{2} g_{ik} R + \lambda g_{ik} = -\varkappa T_{ik} \,.$$

If one is to build more general theories of gravitation than the one just referred to, he will have to make sure that the general Einsteinian covariance leeds merely to trivial identities by using a sufficient number of parameters. In a generalized theory like that, the field equations will be independent from the principle of strong equivalence, hence the dynamical theorems will not follow from the equations of the gravitational field, they will be consequences of the principle of weak equivalence and the laws of the field of matter.

Postulating weak equivalence is enough to get automatically these theorems as identities from the field equations of matter.

6. MORE ABOUT GRAVITATIONAL ENERGY AND RADIATION

6.1. Infinitesimal Coordinate Transformations

An infinitesimal change of coordinates

$$y'^i = y^i + \varepsilon \xi^i + O(\varepsilon^2)$$

with

$$a'^i_k = \delta^i_k + \varepsilon \xi^i_{,k} + O(\varepsilon^2)$$

$$a^i_k = \delta^i_k - \varepsilon \xi^i_{,k} + O(\varepsilon^2)$$

replaces the g^{lm} by

$$g'^{ik}(y') = g^{lm}(y)\, a'^i_l a'^k_m = g^{ik}(y) + \varepsilon\, (g^{im}\xi^k_{,m} + g^{kl}\xi^i_{,l}) + O(\varepsilon^2)\,.$$

But, on the other hand, since the transformation is infinitesimal

$$g'^{ik}(y') = g'^{ik}(y) + \varepsilon g'^{ik}_{,l}\xi^l + O(\varepsilon^2)$$

or

$$g'^{ik}(y) = g^{ik}(y) + \varepsilon\, (g^{im}\xi^k_{,m} + g^{km}\xi^i_{,m} - g^{ik}_{,l}\xi^l) + O(\varepsilon^2)\,.$$

Replacing the partial derivations by covariant ones and applying the relation

$$g^{ik}_{,l} = -\Gamma^i_{ml} g^{mk} - \Gamma^k_{ml} g^{im}$$

we get the variations of the g^{ik}

$$g'^{ik}(y) = g^{ik}(y) + \varepsilon\, (\xi^{i;k} + \xi^{k;i}) + O(\varepsilon^2)$$

and in the same manner

$$g'_{ik}(y) = g_{ik}(y) - \varepsilon\,(\xi_{i;k} + \xi_{k;i})\,.$$

Thus we can always write

$$g'_{il} g'^{lm} \simeq \delta_i^m$$

(see 3.2, 3.1).

6.2. Covariant Mechanics, Killing-Vectors and the Existence of an Energy-Momentum-Vector

Differential identities obtained from the action integral on the ground of its independence from the choice of coordinates constitute what is known as Noether's theorem. If such identities are to bear the significance of conservation theorems for physical magnitudes, the invariance towards coordinate-transformations must correspond to real symmetries of the space-time manifold.

Two kinds of transformations are to be distinguished. Passive transformations are such, which leave bodies and fields unchanged where they are in space-time, whereas the values of the parameters defining coordinates of points are changed. Active transformations leave the coordinates unchanged, whereas the position of bodies and the values of fields are changed.

Passive operations are always possible independently from the structure of space-time. Active ones, on the contrary, may not produce any distortion of that structure and are therefore restructed by the existence of symmetries (isotropy and homogenity) of space-time.

Within the theory of special relativity, Lorentz-Poincaré-transformations of coordinates correspond automatically to the symmetries of space-time. Hence the Poincaré-group describes at the same time possible motions (without distortions) in the Minkowski-space.

The Riemannian space has quite other properties than the Minkowski-space with regards to possible motions: it is much more inflexible. Now, a motion is made out of specific infinitesimal changes of the positions which can be written in the form

$$\bar{y}^k = y^k + \varepsilon \xi^k(y)\,.$$

Having the field of the active operation of motions given by ξ^k, we can then define with it a coordinate-transformation by a passive transformation replacing y^k by

$$y'^k = y^k + \varepsilon\xi^k \quad (\varepsilon \text{ is an infinitesimal magnitude}). \tag{1}$$

This lets a point $Q' = (y')$ correspond to the former point $Q = (y)$. Q' had the coordinates $y^k + \varepsilon\xi^k$ in the old coordinate-system, and it has the coordinates y'^k in the new one.

In the new coordinate-system, the metric at the point Q is given in terms of the new coordinates by

$$\bar{g}_{ik}(Q) = g_{ik}(Q) - \varepsilon g_{lk}\xi^l_{,i} - \varepsilon g_{il}\xi^l_{,k}\,.$$

(We write it $\bar{g}$, for it follows from the motions defined by the field ξ^l.) It can be compared with the metric at Q' in the old system of coordinates

$$g_{ik}(Q') = g_{ik}(Q) + \varepsilon\xi^l g_{ik,l}\,.$$

(We write it g, for it follows from the changes $g_{ik,l}$ in the metric.)

Equ. (1) is admissable as a motion if and only if the differences $\bar{g}_{ik}(Q) - g_{ik}(Q)$ and $g_{ik}(Q') - g_{ik}(Q)$ coincide, for it is then an isometric representation of space-time upon itself. The condition for this to be the case is that

$$\xi_{i;k} + \xi_{k;i} = 0 \tag{2}$$

(see sec. 6.1).

Equ. (2) defines a group of transformations called Lie-group; their solutions are vectors ξ_i called Killing-vectors; the number of such independent solutions is also the number of parameters of the Lie-group.

If space is Minkowskian, there are ten linearly independent Killing-vectors, the group of motions is isomorphic to the inhomogeneous Lorentz-group.

Killing-vectors allow us to replace dynamical equations by integral theorems of conservation. Indeed, from

$$T^{ik}{}_{;k} = 0 \quad \text{and} \quad T^{ik} = T^{ki}$$

follows by (2) that

$$(T^{ik}\xi_k)_{;i} = 0 \quad \text{and} \quad (\mathfrak{T}^{ik}\xi_k)_{,i} \equiv (\sqrt{-g}\,T^{ik}\xi_k)_{,i} = 0\,.$$

Hence

$$\int_{y^0=\text{const.}} \mathfrak{x}\xi_k \mathrm{d}y^1\mathrm{d}y^2\mathrm{d}y^3=\text{const.} \tag{3}$$

independently from the choice of the three-dimensional hypersurface $y^0=$const. on which the integration is performed.

This shows that each Killing-vector ξ^k defines a constant finite magnitude, i.e. allows to establish a theorem of conservation. Minkowski-space, as well as spaces of constant curvature, allows ten such independent conserved components; thus is the maximum attainable for their number; they correspond to the well-known conserved magnitudes in mechanics.

In the most general case of a Riemannian space, one finds no Killing-vectors at all, and no magnitudes are found to be conserved. In stationary (Riemannian) spaces, there is one Killing-vector; it generates conservation of energy.

The connection between invariant conservation theorems and the existence of space-time symmetries refers of course only to those conservation theorems which can be deduced from the invariance of the variational principle with regard to passive transformations of space-time; these are in classical mechanics the theorems of energy-momentum (four), of angular momentum (three), and of the center of mass (three): total ten. Identities deduced as conservation theorems from symmetries which are not space-time symmetries, e.g. conservation of electric charge, keep in the Riemannian manifold the significance they had in the Minkowskian one independently from the modifications of the structure of the manifold. Their differential form is obtained by generally covariant transposition, as is required by the principles of relativity and of equivalence, like in the case of the field equations or the equations of motion (see 4.2).

For instance, if in special relativity theory such a theorem takes the form

$$I^l{}_{,l}=0\ ,$$

allowing for the integral conservation of a scalar quantity (charge, number of particles, etc.)

$$N \sim \int J^0 \mathrm{d}^3x=\text{const.}\ ,$$

it will be written in general relativity theory like this:

$$J^l_{,l} \equiv (\sqrt{-g}I^l)_{,l} = 0 .$$

The same assumptions translated from special to general relativity (Gauss' theorem etc.), will allow to write

$$\int \sqrt{-g}I^l \mathrm{d}S_l = 0 \Rightarrow N \sim \int \sqrt{-g}I^0 \mathrm{d}^3y = \text{const.} ,$$

showing that N is a constant independent of time.

6.3. Energy-Momentum-Vector from Einstein's Identities

Einstein's identities (3.2.4), written e.g. for the material tensor M_i^K (9.2.3):

$$(\mathfrak{M}_i^k + t_i^k)_{,k} = 0$$

lead to integral conservation theorems and define a spatio-temporal Lorentzian vector. By means of Gauss' theorem, the integration yields firstly

$$\oint_S (\mathfrak{M}_i^k + t_i^k) \, \mathrm{d}S_k = 0 ,$$

where S is a closed three-dimensional hypersurface embeded in the four-dimensional space-time V_4. This hypersurface S can be thought of as consisting of one time-like cylindric hypersurface Z and two space-like hypersurfaces V and V' describing a three-dimensional domain at times $y^0 = a$ and $y^0 = b$:

$$S \text{ "=" } Z + V + V' .$$

We compute the integration within an infinite V_3 and request that at space-like infinity, one gets sufficiently soon a Galilean asymptotic structure of the form:

$$g_{ik} \to \eta_{ik} + O\left(\frac{1}{r}\right), \quad g_{\lambda_0} \to O\left(\frac{1}{r^2}\right), \quad g_{ik,l} \to O\left(\frac{1}{r^2}\right)$$

with

$$r \to \infty; \quad \lambda = 1, 2, 3; \quad j, k, l = 0, 1, 2, 3; \quad \eta_{ik} \equiv \overset{\circ}{g}_{ik} .$$

Under these assumptions, the integrals on the time-like hypersurface Z vanish and we are left with

$$\int_{x^0=a} (\mathfrak{M}_i^0 + t_i^0)\, \mathrm{d}V_3 = \int_{x^0=b} (\mathfrak{M}_i^0 + t_i^0)\, \mathrm{d}V_3 .$$

But since the space-like hypersurfaces $x^0 = a$ and $x^0 = b$ are arbitrary, the integral

$$P_i \equiv \int (\mathfrak{M}_i^0 + t_i^0)\, \mathrm{d}V_3$$

itself fulfils two conditions

(i) it is temporally constant $\dfrac{\mathrm{d}P_i}{\mathrm{d}\tau} = 0$,

(ii) it is invariant towards any coordinate-transformation which leaves the limit conditions unchanged.

Clearly, these integrals behave like vectors during Lorentztransformations. Hence Einstein's conservation identities imply the existence of a temporally constant Lorentzian quasi-vector P_i. This vector is called energy-momentum-vector of the system "matter + gravitation".

By taking Einstein's field equations into account, Einstein's identities can also be written

$$\Theta_{i,k}^k \equiv \left(R_i^k - \frac{1}{2}\,\delta_i^k R - t_i^k\right)_{,k} = 0 .$$

Since they are identities, they imply the closing property of a differential form; therefore a quantity ϑ_i^{kl} must exist such that

$$\mathfrak{M}_i^k + t_i^k = \vartheta_{i,l}^{kl}$$

where, by the way,

$$\vartheta_i^{kl} = -\vartheta_i^{lk}.$$

This quantity has been called superpotential by v. Freud. It is an affine tensor density, which depends only upon the metric g_{ik} and the connection Γ_{kl}^i and has been found equal to

$$2\vartheta_i^{kl} = \sqrt{-g}\,[\delta_i^k (g^{rs}\Gamma_{rs}^l - g^{lr}\Gamma_{rs}^s) + \delta_i^l (g^{kr}\Gamma_{rs}^s - g^{rs}\Gamma_{rs}^k)$$
$$+ g^{lr}\Gamma_{ir}^k - g^{kr}\Gamma_{ir}^l] .$$

The vector P_i can be expressed by means of the superpotential, after applying Gauss' theorem, like this

$$P_i = \oint_F \vartheta_i^{\varrho 0} n_\varrho \mathrm{d}F$$

where F is a big two-dimensional sphere of radius $r \to \infty$ the normal on its surface, and $\varrho = 1, 2, 3$. Thus, the energy-momentum-vector P_i of "matter + gravitation" depends, on the ground of Einstein's field equations, only upon the intensity of the gravitational field at infinity; more specially, because of the Galilean limit conditions assumed, only upon the pole-like (source-like, quasi-Newtonian) terms in the gravitational field. This is in complete analogy with the case of Maxwell's theory, in which the field is Coulombian at infinity.

The case of time-independent, so-called static gravitational fields is particularly simple. Beside above limit conditions, it requires that $g_{ik,0} = 0$. For a static centralsymmetric source of mass m, which is an isolated system, the total energy

$$P^0 = \int (\mathfrak{M}_0^0 + t_0^0) \, \mathrm{d}V_3$$

can be computed[1], e.g. for a Schwarzschild solution (see sec. 4.3) and is found equal to mc^2, whereas on the ground of the limit conditions $P^\alpha = 0$ $(\alpha = 1, 2, 3)$. For this case, g_{00} has to be taken equal to

$$g_{00} = 1 - \frac{2Gm}{c^2 r},$$

called quasi-Newtonian potential. The fact that for this case

$$P_k = (mc^2, 0, 0, 0)$$

shows that Galilean limit conditions define the system at rest of a static gravitational field with quasi Newtonian potential as well as of the system at rest of the matter of its source.

6.4. On Gravitational Shock Waves and Plane Waves

On the ground of general covariance, the ten Einstein field equations leave four functions $\varphi^i(y)$, undetermined; hence only six of these equations are independent. The symmetric Ricci-tensor, however, as well as

[1] See e.g. R. C. Tolman, Relativity, Thermodynamics and Cosmology, Oxford at the Clarendon Press 1934, sec. 91, p. 232 ss.

the metric, have each ten algebraically independent components. Therefore, there must exist four differential identities between the R_{ik}, which, invoking Bianchi, assume the form (2.12):

$$G^k_{i,k}+G^m_i\Gamma^l_{ml}-G^l_k\Gamma^k_{il}=0 \tag{1}$$

by using preferably Einstein's tensor G_{ik} ($i, k=0, 1, 2, 3$). Our problem is to construct a Cauchy-problem whereby, e.g., at time $y^0=ct=0$ values will be prescribed. To render this manipulation tractable, let us separate in Bianchi's conditions the derivatives with respect to time $y^0=ct$ and sum up the remaining terms on the right-hand side abridged B_k:

$$G^0_{k,0}=-G^p_{k,p}+G^n_m\Gamma^m_{k,n}-G^n_k\Gamma^l_{nl}\equiv B_k\,. \tag{2}$$

The equations (2) are identities where B_k does not contain derivatives with respect to time higher than to the second order; the same must hold for the left-hand side; hence

$$G^0_k=R^0_k-\frac{1}{2}\,\delta^0_kR \tag{3}$$

does not contain second-order time derivatives. A further examination would show that no second-order time derivatives of g_0 and g_{00} are to be found in the whole relation. Thus, G^0_k contains no time-derivative of g_{k0} whatever.

Furthermore, Einstein's field equations are in involution,[1] meaning that if the six space-space components or the Ricci-tensor vanish everywhere in a domain

$$R_{\mu\nu}=0 \quad (\mu, \nu=1, 2, 3)$$

as well as the four right-hand sides of Ricci's identities at time $t=0$, then Einstein's field equations are verified in the whole domain. This implies that if $G^0_k=0$ at some fixed time, it remains so at all times.

We can therefore use the equation $G^0_k=0$ in order to determine algebraically g_{k0}, and then eliminate g_{k0} from the six remaining field equations, which will then contain only the six functions $g_{\mu\nu}$ ($\mu, \nu=1, 2, 3$) besides first and second derivatives. And since general covariance allows to choose arbitrarily four functions, let us select them in such

[1] Involution in the sense in which Cartan used the term.

a way as to get rid of four suitable combinations of the $g_{\mu\nu}$: In short, $G_k^0=0$, eliminate the four g_{k0} and by an appropriate choice of the coordinates, eliminate four combinations of the $g_{\mu\nu}$; only two essential field functions remain. If they are determined by Einstein's field equations, then the whole field is determined, too.

The main argument will now be to construct these two essential functions (Bondi, 1962).

To illustrate this, let D be a domain of space-time V_4 and let a hypersurface Σ in D be defined by the equation

$$\Sigma_{\text{Def.}}: Z=0$$

separating D into two partial domains:

$$D_+, \quad \text{where} \quad Z>0\,,$$
$$D_-, \quad \text{where} \quad Z<0\,.$$

We shall assume for the sake of simplicity that the $g_{ik}(y)$ are everywhere holomorphic functions in D except on Σ, where the transversal derivatives of a certain order n and higher might be discontinuous (Treder, 1962):

$$\left[\frac{\partial^m}{\partial Z^m} g_{ik}\right] \equiv \left(\frac{\partial^m}{\partial Z^m} g_{ik}\right)_{Z\to+0} - \left(\frac{\partial^m}{\partial Z^m} g_{ik}\right)_{Z\to-0} \neq 0 \quad \text{for} \quad m \geqq n\,. \tag{4}$$

Nevertheless, the field equations $R_{ik}=0$ for the vacuum are supposed to apply everywhere. Such discontinuities of the derivatives of the metric must exist, if the gravitational field is to be modulated so as to propagate information; Σ is thus a wave front.

We shall now show that such new functions really do not exist unless Σ is an isotropic hypersurface. In order to see this, let us, however, develop g_{ik} by adapting the development to the situation, i.e., to the existence of the wave front:

$$g_{ik} = g_{ik}^- + \underset{(n)}{\gamma_{ik}} \underset{(n)}{H}(Z) + \underset{(n+1)}{\gamma_{ik}} \underset{(n+1)}{H}(Z) + \dots \tag{5}$$

Here, g_{ik}^- designates the metric in the partial domain D_-, where $Z<0$ as well as its analytical continuation in the domain D_+; the functions $\underset{(m)}{\gamma_{ik}}$ may very well be holomorphic, whereas $\underset{(m)}{H}(Z)$ denotes discontinuous functions by passing through Σ (Heaviside-function)

$$\underset{(m)}{H}(Z) = \begin{cases} \dfrac{Z^m}{m!} & \text{for} \quad Z \geqq 0 \\ 0 & \text{for} \quad Z<0 \end{cases} \quad \text{for all} \quad m \geqq n\,.$$

In the lowest order of the discontinuity, we obtain

$$(\underset{(n)}{\gamma_{ik}})_{Z=0}=\left[\frac{\partial^n}{\partial Z^n}\,g_{ik}\right].$$

A corresponding development of the Ricci tensor R_{ik} will indeed exist. Since Einstein's field equations are valid throughout D, the coefficients of this latter sequence must all vanish, in order to warrant $R_{ik}=0$:

$$R_{ik}=\frac{\partial}{\partial Z}\,R_{ik}=\ldots=\frac{\partial^m}{\partial Z^m}\,R_{ik}=\ldots=0 \tag{6}$$

whence

$$[R_{ik}]=\left[\frac{\partial}{\partial Z}\,R_{ik}\right]=\left[\frac{\partial^m}{\partial Z^m}\,R_{ik}\right]=\ldots=0\,. \tag{7}$$

The first of the latter equations not to be identically satisfied is—by reason of R_{ik} containing second derivatives of the metric—

$$\left[\frac{\partial^{n-2}}{\partial Z^{n-2}}\,R_{ik}\right]=0\,. \tag{8}$$

According to (8) the $\underset{(n)}{\gamma_{ik}}$ are unequivocally conditioned or specified.

Now, if Σ is either timelike or spacelike-meaning that the normal vector p_i satisfies the inequality

$$g^{ik}p_ip_k \neq 0, \quad p_i=Z_{,i}$$

then equation (8) contains the following six algebraically independent relations:

$$p^lp_l\underset{(n)}{\gamma_{ik}}+g^{ml}\underset{(n)}{\gamma_{ml}}p_ip_k-\underset{(n)}{\gamma_{kl}}p^lp_i=0\,. \tag{9}$$

The general solution $\underset{(n)}{\gamma_{ik}}$ depends consequently upon four functions and can be put as

$$\underset{(n)}{\gamma_{ik}}=\underset{(n)}{a_i}p_k+\underset{(n)}{a_k}p_i\,. \tag{10}$$

But then the first discontinuity in the development of the metric can be eliminated by changing the coordinates to

$$\bar{y}^i=y^i+\underset{(n+1)}{a^i}H(Z)+\ldots \tag{11}$$

(see 6.1) and the next discontinuity takes the place of the eliminated one. However, the same reasoning applies to the new one, and so forth. Hence, discontinuities of that sort are only due to fortuitous choices of coordinates and can be got rid of by changing them. Such fortuitous discontinuities are not essential, they are mere appearances which can be removed and cannot, of course, furnish any real information.

The case of lightlike (isotropic) hypersurfaces is quite different, i.e., if

$$g^{ik}p_ip_k=0$$

for then equation (8) yields four linear algebraic conditions for the $\underset{(n)}{\gamma_{ik}}$ on Σ:

$$-\underset{(n)}{\gamma_{ik}}p^k+\frac{1}{2}\underset{(n)}{\gamma_{kl}}g^{kl}p_i=0\,. \tag{12}$$

From equation (12) it follows that the $\underset{(n)}{\gamma_{ik}}$ are found to consist (i) of a term which again can be transformed through change of coordinates to the elimination of a fortuitous discontinuity, as well as (ii) of a second part which depends upon two functions, say $\underset{(n)}{E}$ and $\underset{(n)}{\delta}$. Hence, the discontinuity is found to be perpendicular to the propagation vector p_i,

$$\underset{(n)}{\gamma_{ik}}=A_ip_k+A_kp_i+\pi_i\pi_k+\tilde{\pi}_i\pi_k \tag{13}$$

where

$$\pi_kp^k=\tilde{\pi}_kp^k=\pi_k\tilde{\pi}^k=p^kp_k=0$$

the π_k are given by Stellmacher's Ansatz (Stellmacher, 1938):

$$\underset{(n)}{\pi_k}=\left(\frac{1}{2}\right)^{\frac{1}{4}}E^{\frac{1}{4}}\left(e^{\overset{i\delta}{(n)}}\omega_k+e^{\overset{-i\delta}{(n)}}\omega_k^*\right), \tag{14a}$$

$$\underset{(n)}{\tilde{\pi}_k}=\left(\frac{1}{2}\right)^{\frac{1}{4}}E^{\frac{1}{4}}\left(e^{\overset{i\delta}{(n)}}\omega_k-e^{\overset{-i\delta}{(n)}}\omega_k^*\right). \tag{14b}$$

Therefore, on isotropic hypersurfaces, essential discontinuities can indeed exist, depending on two further parameters. These discontinuities are the so-called shock waves!

Although the relations (8) are, at first sight, conditions only on Σ, nevertheless the $\underset{(n)}{\gamma_{ik}}$ can be required to fulfil all conditions in the whole space G, implying the relations

$$\left[\frac{\partial^{n-1}}{\partial z^{n-1}} R_{ik}\right]=0\,. \tag{15}$$

Four of these ten equations are then identically satisfied, whereas two equations are homogeneous linear differential equations for the quantities $\underset{(n)}{E}$ and $\underset{(n)}{\delta}$.

Their propagation along the null geodesics that generate the hypersurface Σ is determined by

$$\frac{\partial}{\partial x^l}(\sqrt{-g}\,\underset{(n)}{E}p^l)=0\,, \tag{16}$$

$$\underset{(n)}{\delta}{}_{,l}p^l=\frac{d}{dt}\underset{(n)}{\delta}=0 \tag{17}$$

where

$$p^l p_l=0, \quad p^i{}_{;l}p^l=0\,.$$

The four remaining equations yield now the structure of inhomogeneous algebraic equations allowing us to determine the $\underset{(n+1)}{\gamma_{ik}}$. Therefore, $\underset{(n+1)}{\gamma_{ik}}$ reveal a homogeneous part of the same algebraic structure as the $\underset{(n)}{\gamma_{ik}}$. Hence the $\underset{(n+1)}{\gamma_{ik}}$ depend, in the same sense, on two functions $\underset{(n+1)}{E}$ and $\underset{(n+1)}{\delta}$. With the requirement that these equations on Σ are valid for the whole G, we are able to continue this procedure, getting in every case the functions $\underset{(m)}{E}$ and $\underset{(m)}{\delta}$ with the corresponding linear differential equations, which determine explicitly the propagation along the characteristic rays. (The result obtained for the order $n+1$ is independent of the choice of $\underset{(m)}{\gamma_{ik}}$ $(m>n+1)$ in G.)[1]

The field $g^+_{ik}-g^-_{ik}$ which is of the character of a gravitational disturbance (shock waves) in G depends on $\underset{(m)}{E}$ and $\underset{(m)}{\delta}$, each of these sets forming a so-called "news-function"; their initial data can be given, in an arbitrary way, on the two-dimensional section of the isotropic hypersurface Σ with

[1] H.-J. Treder, Gravitative Stoßwellen, Akademie-Verlag, Berlin 1962.

a spacelike cone $x^0 = \text{const}$. On the total Σ these functions are defined by the propagation equation, and, corresponding to the isotropy of Σ, the disturbances are propagated in V_3 with the local velocity of light. The discontinuity of lowest order is interpreted in a geometrico-optical way, $\underset{(n)}{E}$ being the intensity and $\underset{(n)}{\delta}$ the polarization angle of the propagated disturbance, which shows some similarity with the laws of geometrical optics.

The above analysis proves that the Einsteinian equations admit of a modulation in the gravitational field with information propagating with the velocity of light corresponding to the isotropy of the wave front Σ. The waves are purely transversal–a fact that determines them completely by giving an amplitude and a polarization angle as well as the wave front and the direction of propagation. All these quantities can be furnished in terms of a lightlike coordinate Z, since the choices of $\underset{(m)}{E}$ and $\underset{(m)}{\delta}$ are independently made.

A Taylor-expansion (equation (5)) of g_{ik} induces an analogous series for the Riemannian tensor

$$R_{iklm} = R^{-}_{iklm} + \underset{(n-2)}{P_{iklm}} H(Z) + \ldots$$

R^{-}_{iklm} is defined as the Riemannian tensor for $Z < 0$ or is conceived of as being an analytical continuation into the domain $Z > 0$. The coefficients P_{iklm} are holomorphic functions, the lowest one satisfying the algebraic relations

$$\underset{(n-2)}{P_{iklm}} p^m = 0 \,, \tag{18}$$

$$\varepsilon^{iklm} \underset{(m-2)}{P_{stlm}} p_i = 0 \tag{19}$$

where ε is Levi-Civita's symbol. Conversely, these equations render it possible to define the existence of a gravitational wave front Σ.

Let us finally consider the propagation of a gravitational wave in the flat Minkowski-space E_4. In the simplest case, a wave front exists on a Minkowskian null hypersurface $Z = 0$, the space being flat on the one side ($R^{-}_{iklm} = 0$) with a nonvanishing R_{iklm} on the other side. In the neighbourhood of the wave front as well as on the wave front itself obtains

$$R_{iklm} p^m = 0 \,, \tag{20}$$

$$\varepsilon^{iklm} P_{stlm} p_i = 0 \tag{21}$$

where

$$p^i p_i = g^{ik} p_i p_k = 0 .$$

If these conditions are fulfilled in the whole wave domain, one speaks of gravitational waves with "plane" fronts (Brinkmann, 1923); these waves are indeed "plane gravitational waves", if and only if a coordinate-system can be chosen such as to make the g_{ik} dependent solely upon the isotropic coordinate Z. With $Z \equiv x^0$, we get

$$(g_{ik}) = \begin{pmatrix} 0 & 0 & 0 & 1 \\ 0 & g_{11}(x^0) & g_{12}(x^0) & 0 \\ 0 & g_{12}(x^0) & g_{22}(x^0) & 0 \\ 1 & 0 & 0 & 0 \end{pmatrix} \tag{22a}$$

for which a field equation between g_{11}, g_{12}, and g_{22} reads

$$g^{ik} g_{ik,00} + \frac{1}{2} g^{ik}{}_{,0}\, g_{ik,0} = 0 . \tag{22b}$$

The very question of the real existence (and of experimental evidence) of gravitational radiation is still an open issue. Some physicists are strongly convinced of its very existence. However, the crucial experiments of the "receptions of gravitational waves from the universe" are very controversial. Furthermore, the theoretical arguments for the possibility of an emission of gravitational waves by means of freely moving bodies in a gravitational field are rather bizarre. Free motions in gravitation fields are forceless motions, which we can describe by accelerationfree rest systems in Fermi-coordinates. Free gravitational waves may therefore–according to Infeld–exist from "the first moments of the universe", but there cannot exist sources of gravitational radiations.

In connection with the exact plane gravitational waves Einstein and Rosen (1936) have proved that the form $g_{ik} = g_{ik}(z)$ (see equations (22a) (22b)) is incompatible with the de Donder-condition $(\partial/\partial x^k)\,(-g)^{1/2}\, g^{ik} = 0$ according to the exact equations $R_{ik} = 0$. But the de Donder-condition excludes the nonphysical modes of the plane waves in the linearized theory (4.6). And thus, cogently reasoning, Einstein writes to Max Born: "Ich habe zusammen mit einem jungen Mitarbeiter gefunden, daß es keine Gravitationswellen gibt, trotzdem man dies gemäß der ersten Approximation für sicher hielt" (Einstein, 1936).[1]

A point discussed by Einstein, Lorentz und Schrödinger 1917/18 is: The very tensor which can be interpreted as the symmetrical energy-momentum-tensor of pure gravitational field is the Einstein-tensor. However, this tensor vanishes according to the Einstein field equations.– On the other hand, Einstein's canonical energy-momentum-complex (3.2.4) is an affine-tensor, only, and can be made to be zero on any world-point by a suitable choice of coordinates. Therefore, the energy of gravitational radiation is not a localizable entity!

7. FUTURE OF GRG

7.1. Concepts of Space[2]

When confronted with the world around him, man–if he were completely free of any prejudices–would, it seems, at least feel that there are things in this world, but that they are submitted to changes. The brute facts and his immediate experience cannot but compell him to such a first opinion about the world.

His next reflexion might be, that if there were no changes, then no notice of things would be possibly taken, and if there were no things, then no changes could ever occur. Things are material, changes are temporal: matter and time appear thus so closely entangled, that they, so-to-say, are two aspects of the Same: of the world.

If any problem is to be put–and solved–about this Same, the mind must invent a means to conceptually separate matter and time from each other enought to be able to treat them as distinct, though related, entities: a means which would enable him to accumulate knowledge about this Same, itself made of matter in time and/or material time (or should we say: "real" time, since real comes from Latin *res*: a thing). Such knowledge is then called: Physics.

[1] "I have found, together with a young co-worker, that there are no gravitational waves, in spite of the fact that–in conformity with the first approximation–they were considered a certainty." (translator).

[2] This Part has been published in the Issue dedicated to P. L. Bhatnagar's 60th Birthday of The Mathematical Student (Vol. XL, No. 1, 1972, pp. 49–64).

It was the very great merit of the Greek, among them Euclid, to produce the frame endowed with sufficient structure by means of which this was finally rendered possible. This frame is called (in modern language: three-dimensional, Euclidean) *space*, and the exercise consisting in establishing all its workings is *geometry*.

Therefore, geometry is that, which made a reasonable physics possible, by which the temporal behaviour of matter is analyzed and known. Everybody knows that mathematics grew out of it, together with reflexions on numbers and the algebra. Thus mathematics, especially geometry, became the power with which man is endowed when doing physics as a science. This became so potent, that two thousands years later Descartes–in a sort of Platonic tradition–took it for granted, that geometry "was practically physics" if space were considered to be filled with suitable conceived vortices. His particular conception failed, but Newtonian space proved successful and people came to "believe" in the reality of the Newtonian space: *three*-dimensional and *Euclidean* space, endowed with a *metric*; Pythagoras' theorem applying.

Kant, writing on Pure reason, said that space (and surely he thought of nothing but Newtonian space) was a "form of our sensible intuition a priori". This is very near what we mean when saying that space was invented (by the mind) in order to find out how to solve problems about the distinction between matter and time, though Kant would probably not have accepted the connotation: invented.

Some may contend at this point, that "geo-metry" came out of a quite different preoccupation, viz. the need for land-owners and labourers to handle pieces of arable earth. This is true, too. But we should like to recall that such pieces were estimated in "morning", meaning a piece that could be laboured in a morning's time. This is very much different from a purely metric measure, it is loaded with physics (hardness of the ground, slope etc.), as well perhaps as with other characters. So the reduction of that commerical geometry to pure geometry is of another kind. By the way, it has led to two-dimensional geometry, and the third dimension belonging to it is the direction of gravity, which may be approximately orthogonal in the plain, but never in the mountains. So from there, threefold dimensionality is neither isotropic nor homogeneous.

But it is absolutely correct to abstract pure geometry from natural data in order to get pure mathematical relations. Is then space: form of our intuition, a *real* space? This is doubtful. It is at least an abstraction and, in view of what has been said further above, a means to master the fundamental problem of physics.

Short after Kant's time, Lagrange explained that mechanics can be made more analytical than Newton's by forgetting about Newtonian space and using a space of configuration (q-space). He constructed it with f orthogonal Cartesian coordinates (f degrees of freedom). Hamilton added to it p-space (conjugate momenta), then came phase-space (either Γ-space or μ-space). They are all useful means to solve specific physical problems; they do so in a way quite similar to Newtonian space for classical mechanics. The idea that dynamics is geometry of a q-space was the approach to mechanics chosen by H. Hertz (1894) in his famous book "Die Prinzipien der Mechanik in neuem Zusammenhang dargestellt".

Electrodynamics was not given good care by using this space. Minkowski-space was invented, with two possibilities: 4 real, or 3 real and one imaginary coordinates (with corresponding signatures). Then followed Riemannian space, affine space, Finsler-space etc. along the study of GRG, whereas quantum theory introduced Hilbert-space and general vector spaces with corresponding algebras.

Each time, as we now know, the space used must be construed from a suitable axiomatics. Axioms are not evident truths, they are implicit definitions. Therefore, none of these spaces is "real space". There is no such thing as real space. Yet, many a relativist today might be tempted to say: oh yes, Riemannian space is—at least an approximation to—real space. But why should not a quantum theorist then say: Hilbert space is such an approximation? The answer might be of course: each quantum system needs another Hilbert space, so this is a fiction, whereas the Universe (the totality of what is) needs something like a Riemannian space, it is even identifiable with it. Then we shall ask: with what Riemannian or other space exactly is it identifiable? Give me its metric g_{ik} and all its further properties as final datum, and then everything is determined, is even superdetermined in it; no freedom is left, "time-matter" and "space" are the Same; it is Spinoza's God if you will and we must be pantheists. Practically, of course, we have merely not yet been able to write g_{ik} in terms of suitable coordinates so precisely, that all: matter, life, soul and what have you will automatically be implied.[1]

[1] The energy-momentum-tensor possesses at most ten algebraically independent components, whereas, in order to describe matter completely, one needs

Apart from the uneasiness produced by this eventuality, everyone may have guessed that the kind of revolution which, following the appearance of Einsteinian relativity, has taken place at the beginning of this century, may very well repeat itself: a completely new "geometric" approach may prove necessary in the future, making GRG obsolete and replacing it by some super-theory.

This is a first opening we may expect towards the future. However, it is more than vague, since it rests on nothing but a possible analogy. Still, something like that is pretty sure to happen.

7.1.1. Flatness and Dimensionality

A second consideration refers to Euclidicity. Newtonian space N^3 is Euclidean, we might say: flat. Is this a singular character? No, as far as it can be conceived of as a zero approximation. However, there is a difference, whether one says that the approximation is due to very thin distribution of matter throughout space (which is a physical proposition), or that it is most convenient for mathematical developments as tangent space to a curved one (with is a geometrical proposition), or still that it is bound up with the restriction to e.g. Galilean observers (which is a kinematical proposition).

Minkowski-space is also (pseudo-)Euclidean. Hence, the same alternatives apply. But in flat Minkowski-space, Lorentz-transformations are both passive and active transformations, that is: coordinate transformations, and motions, for the Lorentz-group is at the same time a canonical transformation which can be considered as made up of infinitesimal transformations.

In Riemannian space, these things go apart. In it, passive and active

probably more than these ten terms. This fact, known as Rainich's problem, is really tied up with global and topological structures of space-time, as was noticed by Wheeler. Assuming that all matter appears to be generated by a primary field (as proposed by Heisenberg) endowed with only a few algebraically independent components, we get a unified theory of matter, however not in the same sense as Einstein's. GRG is then implied as a unified theory of space and 'time-matter': Einstein's field equations determine, on the one hand, the metric g_{ik} and the structure of the curvature tensor fields, and on the other hand, they determine explicitly the material tensor of the primary field and hence this field itself (see 5.6).

transformations are comparable if and only if Killing-vectors exist, which needs not be the case necessarily (see 6.2).

Moreover, going over from N^3 to E^4 is not the same as from E^4 to R^4, not only on the ground of the three, resp. the four dimensions in one and in the other cases. From the point of view of the change from N^3 to E^4, Euclidicity, or flatness, might be nevertheless considered as a singular character, because there is matter in N^3 and there is energy (though no other matter) in the electromagnetic vacuum (which is therefore a non-empty vacuum). Yet flatness does, from the point of view of GRG, play the role of an asymptotic condition and was envisaged from the beginning and for quite a time since as a character of emptyness (absence of matter of all kind including energy). It is, after all, true that in any spatial representation matter distribution appears experimentally thin on the whole (apart from condensation nuclei: particles, stars, . . .). So, flatness can in any case be regarded as a very good zero approach. There are –it is true–such as, on the ground of Mach's principle, exclude the "flatness" completely.[1]

Dimensionality is quite another problem. If space were by a naive realist regarded as real, would this space necessarily be e.g. three-dimensional?

As a first answer to that question, we might say: Three-dimensional space allows the existence of knots and laces, two-dimensional space does not. There are knots and laces, e.g. in our bowels. Hence space cannot be two-dimensional. In four-dimensional space, knots can be stretched into straight lines.

A reply to this first answer might consist in saying, that in order to explain a particular aspect of matter, viz. the appearance of what is called knots and laces, we must borrow from geometry a three-dimensional space. But this does not make that space the more real.

A second answer to the same question could consist in saying that the number three might result from the physiology of our eyes: we see

[1] Therefore, according to Einstein's interpretation of Mach's principle–the Mach-Einstein-doctrine–the metric g_{ik} should be completely determined by the matter. The limiting case of a cosmos without matter (and without free gravitation radiation) is not "flat metric" but "non metric", i.e., for empty space $T_{ik} \to 0$ we have not $g_{ik} \to \eta_{ik}$ but $g_{ik} \to 0$ according to the Mach-Einstein-doctrin (cf. H.-J. Treder, Die Relativität der Trägheit, Akademie-Verlag, Berlin 1972).

on both retinae, therefore we notice the depths beside the plane "photographic" images without displacement of the view point; hence, by a transposition from sensorial aperception to the mind, we mean that space is three-dimensional. A witty physicist said once: if man possessed one more eye, which was in motion relative to the rest of his body at a speed approaching c, he would have "known" of Einsteinian (special) relativity long before Galilei. Apart from the difficulties to nourrish that supplementary organ with blood along vessels, the problem is of a different nature.

Space (of the naive realist, or of Newton's) is space; time is time. In popular presentations of special relativity, it has often been said, that time (multiplied by c or by ic) has been made to a fourth dimension of space. Yes in the sense of inventing a new. Minkowskian space as means of consequent description in pure electrodynamics. *No* from the point of view of natural data, i.e. physics: If time and matter are originally entangled as one Same, they are also given, and man as endowed with curiosity and power of investigation does not dispose of them. Especially, when time is made to a parameter by separating it from matter by means of a suitable frame of description, time is given, and we never dispose of it. From this follows that we never measure time proper, we can only retrospectively attribute magnitudes to time intervals from readings on material bodies in the metric of the space used, –or from properties other than lengths or angles. In other words, time is an independent variable. By using Minkowskian space in physics, points in it are called events, world lines are "material" objects, and all observations made on measurable properties must yield numerical values for functions $\varphi(\boldsymbol{x}, t)$ of the four parameters x^0, x^1, x^2, x^3. Therefore x^1, x^2 and x^3 are to be considered like time x^0/c as given independent variables, and relativistic physics must after all, be a physics of fields $\varphi(\boldsymbol{x}, t)$. By the way, time in pre-relativistic quantum theory is a c-number, i.e. not an observable, and relativistic quantum field theory does not alter this fundamentally.

Thus, from that point of view, "time" is made richer in structure: it is the whole of E^4. But at the same time, space as the means of separation between matter and time cannot be the same space as "time" E^4. So, what is it? A possible answer to this is, that if a relativistic theory is necessarily a field theory, then quantum theory is a way of tackling the problem of the separation, using e.g. a Hilbert-space for the purpose.

From this point of view, too, as long as no such thing is done as quantum theory, time and matter are re-united, what the super-determinism of world-lines confirms. Proper time may be the personal time of an observer, this is all what can be done about it, but each observer has its own proper time, whereas the four-dimensional space is a kind of supertime.

In GRG, this is re-inforced. Geometry is on the same ground identical with this richer time. But it is also identical with matter. Hence GRG can be nothing but cosmology: a closed vase.

Should we then necessarily apply a space (suitably invented) as means of tearing this closed vase into pieces?

In a way, the various problems which we have solved in the previous chapters are such pieces: News functions as singularities. Schwarzschild spheres etc.

The problem arises: can we do something more essential? It could be put with regards to the possibility of a conjunction of Relativity and Quantization. It could also be put with regards to the use of a geometry within which supertime will work as parametric. Affine geometry is a means towards the realization of the latter idea; it is not necessary a priori that $\Gamma^i_{kl}=\left\{ {i \atop kl} \right\}$, but by postulating it, a particular conception of "time and matter" is produced; *in that conception, time is of gravitational nature,* for it coincides with gravitational matter; by the way, all matter is gravitational, or in other words gravitation exists whereever there is matter; that is not the case for other interactions. So, perhaps, gravitation is not an interaction, but it is the common feature to all nature to be time-full; however, if gravitation alone is investigated, then nothing particularly temporal is actually found, because it is there implicitly and the whole of GRG properly–as the Cosmology–appears time-less since one cannot make time explicit as long as one does not embed it into a suitably conceived geometrical frame: Here is the greatest paradox of GRG.

This paradoxical situation leads first to two difficult problems, one referring to the conjunction of the ideas of Relativity and Quantization, the other referring to the unification of (other) interactions with gravitation, – as well as to further problems referring to some completely new inventions of space for the purpose of the distinction between time and matter.

7.2. GRG and Quantization

There are two aspects of that subject-matter: Covariant quantization, and Quantization of gravitation.

7.2.1. Covariant Quantization

There have been successive steps in quantum theory: (i) Old quantum theory using the phase integrals $\oint pdq = nh$, (ii) Classical quantum theory based e.g. on Schrödinger's equation $\frac{\hbar}{i}\frac{\partial \Psi}{\partial t} + H\,\psi = 0$, $H = H(p, q, t)$, where t is Newtonian parametric time and where, on the ground of complementarity, it makes no final difference whether q is meant as designating the parametric coordinates or operatorial ones, as long as p designates momenta operators, (iii) Quantum field theory with relativistic Lagrange-Euler-equations built from Lagrange-densities of the field and its derivatives in E^4 accompanied by commutation rules between the eigen-solutions, using relativistic delta-functions; here, t and the q's are parametric only.

A (iv)th step might consist in elaborating a quantum field theory in a covariant way referring to R^4, independently from the consideration of particular interactions. For "free fields", this looks like an academic problem, though important enough since all the mathematical tools must be generalized. This transcription has been undertaken and weighty results have been achieved, mainly by Lichnérowicz. What it does, is to replace Minkowskian supertime by Riemannian supertime within quantum field theory.

At this place, we might call the attention of the reader that time, if it is supertime, i.e. gravitation itself, must be measured in units of the same kind. Such a unit is the *year*, but not the day. The day may be considered by some as a unit of an "inertial time", however, this does not really make sense, because no day could be "observed" without light, and therefore the day is as much an electromagnetic unit of time, yet not related in a simple manner to the unit of time referring to an electromagnetic clock (wrist-watch, atomic clock etc.).

7.2.2. Quantizing Gravitation

To quantize gravitation is something else. One can for example say that, since there are weak field plane waves, their quantization by means of a box attribute to them free particles having the properties of so-called gravitons. This is a de Brogliean point of view. But contrarily to normally conceived particles in quantum physics, gravitons will have to "interact" with one another, because actually Einstein's equations are non-linear and no superposition principle applies. But then it is not easy to make wave-packets, eventually news functions out of the gravitons.

Therefore, quantization should be attempted at without the weak field assumption. A reasonable way to tackle the problem might seem to have to start with a canonical description of GRG, since this is the way quantum theory was translated from pre-quantum mechanics. Dirac, Bergmann and several authors advocate this line of work.

But if the view is accepted that R^4 is fundamentally (four-parametric super-)time, then quantization of gravitation (of gravitation alone) amounts to "quantizing time". So, gravitons, if there are such, would be time quanta impressed upon the material world.

A lot of thinking will still be necessary before such epistemological questions can be suitably put and solved.

7.3. Unification, or the "Generalization" of the General Theory

If gravitation is an interaction like electromagnetism, like the weak or like the strong interaction, then a unification of all interactions on the same level should be attempted. Partial solutions of that problem might be given–if they succeed–by unifying gravitation with only one other interaction, e.g. electromagnetism. This has been attempted by many authors, especially Einstein himself. There are mainly four procedures: the socalled already unified theories, the use of projective geometry and fivedimensional varieties with conditions of cylindricity or the like, the use of Finslerian geometry, the use of affine geometry.

The condition of a good unification would be, that all the characters of electromagnetism (and/or of other interaction) are without exception implicitly contained in the unified theory, meaning that especially Maxwell's equations or similar field equations for the fields responsible of the interaction must follow from the general equations (generalized Einstein equations) of the theory. None of the attempts made so far have been completely satisfactory in this respect.

However and again, if gravitation is not an interaction like the other interactions, but supertime itself, then unification may be an erroneous postulate, or a pseudo-problem.

It is not excluded, by the way, that different interactions are essentially non-unificable.

Then what is GRG?

7.3.1. Mathematical Possibilities within Real Riemannian Four-Dimensional Manifolds

In the frame of differential geometry, several ways can be followed in generalizing Einstein's General Theory.

7.3.1.1. General Lagrangians

Like Newton's equations, Einstein's field equations are quasi-linear differential equations of second order ("quasi-linear" means that they are linear in the highest order). Consequently, the possible solutions will multiply as one considers Lagrangians different from the one chosen originally by Einstein; viz. if beside, or instead of (3.2.3)

$$L_{\text{Einstein}} = \sqrt{-g}\, g^{ik} R_{ik}$$

an expression of the form $L\{R_{iklm}, g_{rs}\}$ dependent upon more general combinations of all the components of the curvature and their covariant derivatives is taken into account. E. G. Weyl has proposed

$$L_{\text{Weyl}} = \sqrt{-g} R^2 + \text{const}_I \sqrt{-g} R_{iklm} R^{iklm} + \text{const}_{II} \sqrt{-g} R_{kl} R^{kl}\,.$$

According to Hilbert, one must put

$$\frac{1}{\sqrt{-g}} \frac{\delta L}{\delta g^{kl}} = -\varkappa T_{kl}$$

in order to obtain the right equation for $T^{kl}_{;l}$.

7.3.1.2. Rainich's Proposal

Instead of dealing with Lagrangians, Schrödinger suggested that Einstein's equations

$$R_{ik} - \frac{1}{2} g_{ik} R = -\varkappa T_{ik}$$

should be considered as the definition of the material tensor T_{ik}. Then, all equations satisfied by Ricci's tensor are conditions for that material tensor. In particular, Rainich (1925) has shown that, except for degenerate cases, the following conditions

$$g^{st} [(R_{sp} R^{sp})^{-1} (R^k_{m;n} R_{kl} + R^k_{n;l} R_{km} + R^k_{l;m} R_{kn})]_{;n} = 0 ,$$

$$R_{ms} R^{ns} = \frac{1}{4} \delta^n_m R_{sr} R^{st} ,$$

$$R = 0$$

imply a material tensor of the form

$$T_{ik} = \frac{1}{4} g_{ik} F_{st} F^{st} - F_{it} F^t_k$$

in which the anti-symmetric tensor $F_{ik} = -F_{ki}$ fulfils precisely Maxwell's equations. In that sense, Rainich's proposal unifies the Einstein-Maxwell field, for it shows how Maxwell's electromagnetic field can be thought of as incorporated in space-time structure.

7.3.2. Five-dimensional Theories

The simplest way to a generalization seems to consist in enlarging once more the number of dimensions, from $3+1$ to $4+1$, keeping the Riemannian character of the manifold V_5. Yet, the fifth dimension needs not bear a physical dimension. Therefore, only special V_5's will be physically significant. viz., such in which by a suitable choice of coordinates the 5th dimension can be eliminated (cylindricity condition). Thus, the Riemannian V_5 becomes a projective four-dimensional manifold and the five-dimensional metrical tensor g_{AB} ($A, B = 0, 1, 2, 3, 4$) splits into a four-dimensional metric and a vector and a scalar field in V_4.

This yields a Riemann-Einstein-manifold endowed with a certain matter tensor which is a function of a vector and a scalar field. Kaluza and Klein have suggested to try unification of the Einsteinian and the Maxwellian fields along that line. It is found that, for a constant scalar field, the five-dimensional Einsteinian vacuum equations $R_{AB}=0$ are equivalent with the four-dimensional Einstein-Maxwell-equations.

7.3.3. Giving up Riemannian Structure as (Unique) Source of Gravitation

A third possibility rests with the enrichment of the structure of V_4. We shall consider four such cases (cf. 2.15).

7.3.3.1. Affinity as Independent Field

In Riemannian manifolds, homothetic curvature R^m_{mkl} and torsion $\frac{1}{2}(\Gamma^i_{kl}-\Gamma^i_{lk})$ vanish identically. But, why should they not be functions of the physical data? The g_{ik} and Γ^i_{kl} can be introduced as independent fields. However, since not all non-Riemannian expressions can in an invariant manner be split into a Riemannian part plus supplementary terms, actually the old Einsteinian theory follows, accompanied by a (possibly unphysical) material field.

7.3.3.2. Finslerian Geometry

Instead of looking for the root of the metric in the metrical spinvectors $\sigma_i^{\mu\nu}$, one can look upon the tetrad-fields h_i^A as fundamental. Through the field equations and the relations (cf. 2.15; 5.3).

$$g_{ik}=h_i^A h_k^B \eta_{AB}\,,$$

the metric gets first its definition. This conception differs from the usual one by the fact that a comparison at distance does not depend only on scalars, but also upon directions. Indeed, the Einsteinian affinity can then be expressed by means of the $h_i^A(\{x^k\})$ like this (2.15; 3.5):

$$\Gamma^i_{kl}=h^i_A h^A_{k,l}=\Delta^i_{kl}$$

where the curvature tensor obtained from the Δ^i_{kl} vanishes identically. The physical constancy of e.g. a vectorial magnitude A^l means then, that

$$A^B_{,l} = (A^i h^B_i)_{,l} = h^B_i \,(A^i_{,l} + \Delta^i_{kl} A^k) = 0 \, .$$

Einstein was the first to suggest such a modification (1928).

A geometry in which metric and affinity are functions of the coordinates as well as of the directions has been elaborated by Finsler. Its systematic use towards the obtention of a unified field theory in which electromagnetism is automatically an integral part of the field equations has been proposed by Schaffhauser and Mercier. Unfortunately, such attempts, if they are satisfactory for vacuum fields, are never so for the sources (point charges), which must be borrowed from the exterior and thus do not belong to the unified picture.

7.3.3.3. Supplementary Scalar Field

In a field theory which includes parallelism at a distance, the field has *a priori* not just 10, but 16 components. It is possible to reduce them to 11, viz. 10 from a symmetric metric g_{ik} plus a supplementary scalar field $\varphi(x)$ to be justified by cosmological (or observational) arguments.

Indeed, our information on the cosmos rests upon the reception of electromagnetic waves. Now, Maxwell's vacuum equations are invariant with regard to conformal transformations of the metric of the kind

$$\bar{g}_{ik} = \varphi g_{ik} \, ,$$

i.e. invariant towards the substitutions

$$\mathrm{d}s^2 = g_{ik} \mathrm{d}x^i \mathrm{d}x^k \rightarrow \mathrm{d}s^{-2} = \bar{g}_{ik} \mathrm{d}x^i \mathrm{d}x^k \, ;$$

therefore, as a matter of fact, the metric ds is univocally determined only by *local* experiments performed with material standards (i.e. with moving particles endowed with rest mass). Thus, concurrent cosmological theories could as well appear in conformally transformed metrics ds^{-2}; the problem is, how to realize their junction with local physics.

It has been proposed to interprete scalar-tensor theories such as suggested further above as theories of gravitation in which the "constant" of gravitation is a field function (Dirac, Jordan, Dicke, ...). According to Pauli, the fixation of the metric has then to be done by requiring the local validity of the principle of weak equivalence.

It should be recalled here that the oldest relativistic theory of gravitation, viz. the one advocated already in 1913 by Nordström, was a kind of degenerate scalar-tensor-theory, since, according to Einstein and Fokker, it implies that the metric tensor of V_4 can always be put into the form

$$g_{ik} = \varphi \eta_{ik}.$$

7.3.3.4. Complex Manifold

Finally, there remains the possibility to replace the real by a complex manifold V_4 in which the (formerly symmetric) metric tensor is no more symmetric, but Hermitian:

$$g^*_{ik} = g_{ki} \leftrightarrow g_{ik} \equiv a_{ik} + i f_{ik} \quad \text{with} \quad a_{ik} = a_{ki}, \quad f_{ik} = -f_{ki}.$$

The affinity must also be assumed Hermitian, as well as other field tensors:

$$\Gamma^i_{kl} = \Gamma^{*i}_{lk}, \quad R_{ik} = R^*_{ki}, \ldots$$

The simplest of such theories is the one proposed in 1945 by Einstein and Straus, in which the Hermitian tensors depend upon real coordinates and consequently the square of the line-element is equal to

$$ds^2 \equiv g_{ik} dx^i dx^k = a_{ik} dx^i dx^k.$$

The Hermiticity of the affinity is then insured by the so-called plus-minus-relation

$$g_{\underline{i}\underline{k};l} = g_{ik,l} - g_{rk}\Gamma^r_{il} - g_{ir}\Gamma^r_{lk} = 0$$

(cf. sec. 2.10), from which, for a symmetric metric $g_{ik} = g_{ki}$, the Christoffel-affinity follows univocally. The Einstein-Straus-equations

$$R_{ik} = 0, \quad \Gamma^r_{ir} - \Gamma^r_{ri} = 2\Gamma_i = 0$$

are then Hermitian and have the property, that for a real metric, they reduce to the ordinary vacuum equations of general relativity.

7.3.4. Interpretation of Such Generalizations

For all these attempts, two interpretations are possible: Either the new magnitudes are considered as further components of the gravitational field, so that this field has more such components than the $10 g_{ik}$, or the new magnitudes are interpreted as components of a non-gravitational field, and the generalized theory serves as a unified theory unifying gravitation with some non-gravitational phenomenon like electricity.

According to Pauli, such attempts at generalizations remain purely formal when they allow for an invariant decomposition of the fundamental magnitudes into a Riemannian part and a non-Riemannian part on the ground of general postulates of symmetry. For, then, the ordinary general theory of relativity is valid beside allowing for a "special material field" which may be unphysical.

Undecomposable generalizations are at hand in the field theory with distant parallelism and the Hermitian theory. It seems natural to interpret the non-decomposable structures of such theories as purely gravitational, because they are universal, and the only interaction known to be universal is precisely gravitation.

More fundamental than the question of the possibility of such generalizations is of course the question of their necessity. Some say, that the latter is completely open, for there appears to be no experimental reason for a correction of the Einsteinian theory. However, this is not quite the right argument, for the question is not, whether Einstein's theory does or does not explain the facts with the best possible accuracy. The question is rather: Are there experimental reasons for a unification under one single theory? To this question, the answer is: yes. The argument runs as follows: There are clocks based on gravitation (Earth around the Sun) with the year as unit, "kitchen" clocks with a pendulum); their period can (conventionally) be declared regular. There are clocks based on electromagnetism (wristwatches, ammonium clocks, quartz clocks ...): they are found to be regular when compared with gravitational clocks. That is even one of the most accurate, though indirect, measurements of physics. Therefore, the two interactions seem to have something in common and call for a "single" explanation. Every physicist "believes" that the same will hold for clocks based on weak or on strong interaction, though the experiments do not yield such an accuracy.

A more theoretical reason for revising the classical relativity theory is found in certain difficulties of bringing it together with the equations for material fields (think of the problem of the collapse (4.8.)).

Another argument can be based on a kind of logical inspection of the whole structure of GRG: How far and how completely do Einstein's principles of covariance, of relativity and of equivalence specify the form of the theory–or are more general expressions of these principles possible? Then, if such modifications of the theory of relativity are put forward, what are their physical consequences?

The fundamental notion of Einstein's theory is Riemann's concept of a geodesic measure as implied in the metric g_{ik}. It was Eddington's idea to consider the affine connection Γ^i_{kl} as more fundamental, meaning that, physically speaking, a Foucault's gyroscope is a more elementary measuring apparatus than a rod.

However, Einstein's original foundation of general relativity was based on coordinate-systems. Their mathematics rests on tetrad-fields h^A_i, which in that way became primary to the g_{ik}'s and according to Weyl help to understand better the relation between geometry and matter. Hence, the various generalizations of Einstein's theory of gravitation can be viewed upon as attempts to consider the metric structure of space-time as secondary to more fundamental structures of real physical nature.

In the latter respect, one can regard as adequate the conception according to which physical evolution does not depend on one single, but on two metrical determinations of V_4. Rosen (1940) has suggested to interpret Einstein's equations as field equations defining, in a flat Minkowskian V_4, the g_{ik} belonging–according to the principle of weak equivalence–to phenomenological matter, while these equations themselves should work in that flat V_4. Rosen's interpretation, as will be seen in the next section, has the advantage of facilitating the formulation of a quantization of the gravitational field. However, its disadvantage consists in the fact that, then, Einstein's equations do not, on account of their general covariance, suffice to determine completely the metric, and consequently Rosen's theory requires four *ad hoc* equations which, though they are covariant with regard to the so-called bi-metric, are not so toward the Riemannian metric itself.

7.3.4.1. Are "Time" and "Gravitation" the "Same"?

Finally, the following considerations may be of importance to the question whether generalizations of the theory are desirable in the sense of unifying the various interactions.

It has already been noticed that, among all interactions, gravitation is the only universal one: Electromagnetic forces act only where there are charges, their presence being responsible for the electric part of the field, their motion for its magnetic part; these "forces" are propagated by photons. Strong interaction requires the presence of baryons, and is propagated by mesons; weak interaction is propagated by leptons. If gravitation is a comparable phenomenon, it is propagated by gràvitons; however, gravitons themselves will have to exchange gravitons, because gravitation is "universal"; this fact is related also to the non-linearity of the equations of gravitation. Now, gravitation itself is interpreted as the structure of the Riemannian variety of a manifold V_4. This variety is itself a modification of a flat "Minkowskian" one which would apply if there were no gravitation and which replaced and made obsolete Newtonian space different from time. But the passage from Newtonian description with Galilean relativity to Minkowskian description (and then to Riemannian description) does not mean that *time* is being made a fourth dimension of a physical space endowed with independent reality; much rather it means that, instead of physics consisting in looking for (or predicting) functions $f(t)$ of time t for all physical properties of bodies and systems, physics consists in looking for functions $\varphi(x^0x^1x^2x^3)$ of four *time-like* variables x^0, x^1, x^2, x^3, becoming consequently a physics of fields φ (eventually tensorial, or spinorial in character) submitted to Lorentz-covariance in the absence of gravitation and to general covariance in its presence. However, a flat space is an "empty" space, meaning void of anything capable of curving it into a Riemannian manifold, and for this reason, though we do not mean to attribute any physical reality to either flat or nonflat space-time, the flat "empty" space-time is of no other interest than allowing for the definition of "free fields", i.e. such fields as, even if they may be quantized by some rules borrowed from quantum field theory, will never enter any interaction as long as they remain free. Therefore, only a curved space-time makes itself noticeable as the possible frame-work having the nature of independent variables x^0, x^1, x^2, x^3 and upon which fields may depend; but since *this nature* is *on the one hand* the nature of *time* as of "the given" or imposed (in contradistinction to "the looked for" or predictable), and it is on the other hand the nature of *gravitation* as the Riemannian structure of the manifold through which these variables run in all directions like the threads of a net, the nature of time and the nature of gravitation appear to be one single nature. In short (and inspite of the surprising result of the argument) one might say, that time and gravitation are "the same", for, without gravitation there would be no notice of time—meant: of space-time—on the ground of its quasi-"absolute" isotropy and homogeneity as expressed by the flatness of the Minkowskian (quasi-Euclidean) variety of the empty manifold, and, if there were not the manifold called (space-)time as endowed with sufficient structure, no matter would be noticeable for it must show up fluctuations (in, or *of* space-time) if it is at all to be noticed, and these fluctuations are gravitation itself.

From the latter point of view, gravitation is *not* an interaction; it is time itself. But gravitation is also matter itself as you prefer, for Einstein's field equations equate it with matter, and so time and matter are one. This *one* is *physical reality*. However, since the prime necessity of physics is to distinguish something called Time from something called Matter, a procedure must be invented in order to separate them into two concepts, else no prediction of any sort can be made. The history of physics shows that the procedure always consists in the *in*vention of a geometric structure then used as a *con*vention in the sense of the acceptation of rules of reasoning. Newton's use of a three-dimensional space was a successful but insufficient attempt, based on the convention that the space invented by Euclid is the correct one. Its insufficiency was not clearly recognized until much later, however already Newton's contemporaries felt the danger which it implied, viz. action at distance with infinite velocity. The Einsteinian description of gravitation does not suffer from the same insufficiency because the Riemannian manifold possesses a recognizable *double nature*: it is *space-time as distinguished from matter-gravitation*, but it is also the *procedure of separation* between time and matter.

A primary question immediately arises: How does it effectuate the separation? The answer to this first question is: through the affine connection. A second question arises: In what does the distinction between time and matter root? The answer to this is: in the possession of its particular metrics, including the pseudo-Euclidicity of infinitesimal domains ensuring the possibility to "take one's time" in physical measurements on the ground of the finitude of a universal velocity (falsely conceived of as being primarily the velocity of light). But precisely the fact that the affine connection and the Christoffel symbol coincide: $\Gamma^i_{kl} = \left\{ {i \atop kl} \right\}$ show that time (separated from matter by the Γ's) and gravitation (its field g_{ik}, hence the { }'s) are one and universal at the same time, and that gravitation is therefore not an interaction proper like the ("other") interactions.

These are epistemological considerations. They initiate speculations as to the possible understanding of physics, as distinguished from physics itself. It is well known that Einstein attached much importance to epistemology. His position, which culminated in his controversy with Niels Bohr, was that of a re-inforcement of determinism. Good physics, he said, should not accept complementarity as a necessary principle, for nothing should be by principle left out of the set of measurable properties of systems in favour of something else during one and the same observation, as is the case from de Broglie's duality or from Heisenberg's indeterminacy. Indeed, Einstein's theory of gravitation does not drop a single property, much rather it tightens everything so closely together that the only consequent epistemological interpretation of GRG is that of a superdeterminism. This is supported also by the epistemological argument leading to above equivalence between gravitation and time as the two aspects of what is physically universal.

But time is not a measurable quantity, since it is that, which is given, which is therefore not disposable of. This is expressed in quantum physics by saying that time is a c-number, i.e. not an observable, not a q-number. For this reason, and in the light of epistemological considerations, one must expect that every attempt at a "quantization of gravitation" will encounter the greatest difficulties. Of course, all this assumes the acceptance of Einstein's conception of gravitation. So let us have a last look at the attempts which have been made towards its quantization.

A discussion of the possible relation between GRG and Quantum Theory does not merely arise from the wishes of having either quantization applied to gravitation or covariance applied to quantum fields (or both), but—as may be understandable from above considerations—also from the feeling that superdeterminism leaves no room for a local physics and yet we are local physicists who must (or will) reckon with information obtained at our location in space-time on what was originally alien to us.

7.4. GRG and Quantum Field Theory

Quantum field theory requires, that the matter tensor T_{ik} appearing on the right-hand side of the equations of gravitation is a field-operator if fields are quantized. Consequently, Einstein's tensor $R_{ik}-\frac{1}{2}g_{ik}R$ must be a field operator, too. This assumes of course, that the gravitational field is a tensorial field with quanta (gravitons) of spin 2. (Covariance makes vectorial and scalar gravitons impossible.)

Quantization of the gravitational field can be achieved by canonical methods applied to a degenerate field (i.e. under constraints). A technical difficulty resides in the gravitational equations not being linear, which requires that causality functions themselves (like propagators) be made into operators.

The following remark must be made: Commutators, causality functions and the like are defined in space-time; but space-time itself is determined by the g_{ik}, i.e. by the field of gravitation. Hence a quantization of gravitation implies a quantization of the metric structure of space-time. On the other hand, spatio-temporal concepts like line-element, distance or curvature are of the nature of c-numbers. The only way to unify the two characters of space-time concepts is to assume that geometrical magnitudes are the statistical expectation values of corresponding fieldoperators. An essential modification of space-time geometry is therefore necessary if the structure of differential geometry itself requires a statistical interpretation.

Dirac has proposed a consistent scheme for a canonical quantization of Einstein's gravitational equations, which is based on the singularization of a space-like hypersurface $x^0=\text{const.}$, itself a V_3 of metric $g_{\mu\nu}$ in which momenta $p^{\alpha\beta}$ canonically conjugate to the $g_{\alpha\beta}$ are proportional to $\Gamma^0_{\alpha\beta}$. The commutation rules for V_3 are then written as follows

$$[g_{\mu\nu}(x), p^{\alpha\beta}(x')]=\frac{1}{2}(\delta^\alpha_\mu\delta^\beta_\nu+\delta^\alpha_\nu\delta^\beta_\mu)\,\delta\,(x-x')\,,$$

where a Greek index runs from 1 to 3 (coordinates in V_3) and x stands for (x^1, x^2, x^3). As explained further above, the metric $g_{\mu\nu}$ of V_3 must be an operator when used in these rules, and the very concept of space-like-ness needs a new definition.

It is much easier to quantize the linearized theory within a Minkowskian continuous background. It makes the whole thing only technically, not fundamentally, different from the quantization of a linearized perturbation of gravitation in a given Riemannian V_4. Of course, such a linearized theory does not take the whole of Einstein's equations into account. Moreover, the supposed weakness of the gravitational field as resulting from linearization is rendered illusory even within small space-time domains by the quantum-theoretical fluctuations of this very field. For, by Heisenberg's uncertainty relations involving uncertainties ΔL in the location of a point, the order of magnitude of the fluctuations Δg_{ik} and $\Delta \Gamma^i_{kl}$ of the metric and of the affinity are found to enter relations of the form

$$\Delta g \, . \, (\Delta L)^2 \gtrsim hGc^{-3} \equiv l^2$$

$$\Delta \Gamma \, . \, (\Delta L)^3 \gtrsim hGc^{-3} \equiv l^2$$

showing that, when $\Delta L \to 0$, the fluctuations of the gravitational field become arbitrarily large. Hence, the linearized theory loses its very meaning. The fluctuations of the metric g_{ik} are of the order of magnitude $\Delta g \simeq l$ when ΔL itself is of the order of magnitude

$$\Delta L \simeq \sqrt{hGc^{-3}} = l \simeq 10^{-33} \text{ cm} ,$$

a constant called Planck's elementary length, which Planck introduced already 1906 as the constant relating gravitation (by its Newtonian constant G), relativity (as expressed by the limit velocity c) and the quantum h.

Now, Δg must be kept less than unity for the following reason: The presence of the elementary length l rests on the fact that, following the principle of equivalence, the ratio between inertial and grave mass is constant and hence an increase of inertia of a body involved in measurement is accompanied by a corresponding increase of its gravitative influence on the g_{ik}. The geometrical significance of the g_{ik} requires then that their fluctuations Δg_{ik} do not become too large, otherwise the signature of Minkowskian space-time (or also the space-like-ness of Dirac's hypersurface $x^0 = \text{const.}$) might be destroyed. This upper limit is precisely

$$|\Delta g| < l ,$$

from

$$\eta_{ik} = \begin{pmatrix} l & & & \\ & -l & & \\ & & -l & \\ & & & -l \end{pmatrix} .$$

Consequently, the notion of length loses its physical significance below the limit l.

Now, if a measurement of a length L is required, a standard must be used with a de Broglie wave length Λ smaller than

$$\Lambda = hcE^{-l} < L ,$$

where E is its (kinetic) energy $E = cp$, p being its momentum. But this Λ itself cannot be made infinitely small, its lowest limit being the Schwarzschild radius

$$a = GEc^{-4}$$

of the standard considered as a measuring particle of spherical symmetry; otherwise this standard would so-to-say collapse into its own Schwarzschild horizon and become unable to yield any information. Hence

$$a \lesssim \Lambda \lesssim L .$$

Consequently, l is not only the lowest limit of length as far as its physical significance is concerned, but also its absolute lowest limit as far as physical possibility is concerned: no length smaller than l can be measured.

Quantum theory of gravitation therefore seems to indicate that within "pico-physical" domains, neither the concepts of Riemannian geometry can simply be used anymore, nor even concepts borrowed from the geometry of metric manifolds. But this would finally mean that, if a fusion of quantum theory and GRG succeeds, the fusion will have considerable consequences for the very idea of a quantum field theory.

A last remark must finally be made, which applies as soon as strong gravitational fields occur, i.e. where the curvature of V_4 is large or the radius of curvature small: All information about the microcosmos are obtained by macrophysical experiments. In other words, quantum theory as the theory concerned–according to Niels Bohr–with microphysics requires classical macrophysics as a metatheory for its own interpretation. But the possibility to operate on macrophysical scale is conditioned by the existing structure of the domain of space-time in which the operation is to take place. Suppose that in such a domain the deviations due to curvature attain orders of magnitude comparable with the values obtained from the spatial and temporal observations as made necessary by macroscopic measurement. Then operations which were possible in a quasi-flat V_4 become unfeasable in those domains of the true V_4, or at least they lose their original significance. In other words, unaware of the particular influence which gravitation has upon elementary (i.e. microphysical) processes, the macrophysical theory of measurement necessary for the interpretation of elementary (microphysical) magnitudes has to be changed: A critical revision of quantum theory of measurement–as advocated in former days by Niels Bohr–, in the presence of strong gravitation fields, is still to be achieved.

Gravitationstheorie
und Theorie der Elementarteilchen

Wiederabdruck ausgewählter Beiträge des Einstein-Symposiums 1965 in Berlin

Herausgegeben im Auftrage der Einstein-Kommission beim Präsidium der Akademie der Wissenschaften der DDR

von Hans-Jürgen Treder

1979. 128 Seiten – 3 Abbildungen – 8° – 18,– M

Bestell-Nr. 762 605 1 (6506)

Auf dem Einstein-Symposium, das 1965 von der Akademie der Wissenschaften der DDR aus Anlaß des 50. Jahrestages der Begründung der allgemeinen Relativitätstheorie und Gravitationstheorie durch Albert Einstein veranstaltet wurde, hielten einige führende Physiker und Mathematiker Vorträge über Grundlagenfragen der physikalischen Theorie von Raum, Zeit und Schwere. Sie sprachen ferner über die möglichen Beziehungen der Einsteinschen Relativitätstheorie zur Quantentheorie und zur Physik der Elementarteilchen. In diesen Vorträgen wurden auch zukünftige Entwicklungen der Theorie prognostisch vorweggenommen, u. a. in den Vorträgen der inzwischen verstorbenen berühmten Physiker V. A. Fock (Leningrad) und L. Rosenfeld (Kopenhagen), in denen diese ihre grundlegenden Gedanken über die Stellung der Einsteinschen Gravitationstheorie in der Physik und über die Quantentheorie des Gravitationsfeldes abschließend entwickelten.

Der vorliegende Band enthält den Wiederabdruck dieser Grundlagen-Kapitel mit einer neuen Einleitung des Herausgebers, H.-J. Treder.

Albert Einstein

Akademie-Vorträge

Wiederabdruck durch die Akademie der Wissenschaften der DDR

1979. 578 Seiten – gr. 8° – in Leinenkassette eingelegt 75,– M

Bestell-Nr. 762 581 7 (6495)

Die Kassette enthält in photomechanischem Wiederabdruck 47 der von Albert Einstein in den Sitzungen der Preußischen Akademie gehaltenen Vorträge, die der Akademie der Wissenschaften der DDR als Sonderdruck aus den seinerzeitigen Sitzungsberichten der Akademie vorliegen.

Einsteins Akademieabhandlungen enthalten die ersten grundlegenden Mitteilungen über die allgemeine Relativitätstheorie, seine Begründung der relativistischen Kosmologie und die Fundamente der Quantenstatistik. Wegweisend für die heutige Physik waren auch Einsteins zahlreiche Mitteilungen über die einheitliche Feldtheorie und zur Geometrisierung der Physik mit ihrer Neubewertung der Beziehungen zwischen Physik und Mathematik. Andere Akademiemitteilungen Einsteins zeigen ihn als ingeniösen Experimentator und als Anreger von Grundversuchen der Physik.

Mitten in die heutigen Hauptprobleme der physikalischen Forschung führen Einsteins Diskussionen der Grundlagenfragen der Quantenmechanik und der Beziehungen von Quanten und Relativität sowie von Elementarteilchen- und Feldphysik.

Die Akademieschriften Einsteins bestimmten die Geschichte der Physik und sind die großartigsten Dokumente physikalischen Denkens und Forschens.